# OXIDATION

BANTAM.
BALLS 'O FIRE!

HASTILY
OXIDIZING
STEEL WOOL

## TASK CARD SERIES

Conceived and
written by

**RON MARSON**

Illustrated by

**PEG MARSON**

**TOPS** LEARNING SYSTEMS

342 S Plumas Street
Willows, CA 95988

# WHAT CAN YOU COPY?

Dear Educator,

Please honor our copyright restrictions. We offer liberal options and guidelines below with the intention of balancing your needs with ours. When you buy these labs and use them for your own teaching, you sustain our work. If you "loan" or circulate copies to others without compensating TOPS, you squeeze us financially, and make it harder for our small non-profit to survive. Our well-being rests in your hands. Please help us keep our low-cost, creative lessons available to students everywhere. Thank you!

## PURCHASE, ROYALTY and LICENSE OPTIONS

### TEACHERS, HOMESCHOOLERS, LIBRARIES:

We do all we can to keep our prices low. Like any business, we have ongoing expenses to meet. We trust our users to observe the terms of our copyright restrictions. While we prefer that all users purchase their own TOPS labs, we accept that real-life situations sometimes call for flexibility.

*Reselling, trading, or loaning our materials is prohibited* unless one or both parties contribute an Honor System Royalty as fair compensation for value received. We suggest the following amounts – let your conscience be your guide.

*HONOR SYSTEM ROYALTIES:* If making copies from a library, or sharing copies with colleagues, please calculate their value at 50 cents per lesson, or 25 cents for homeschoolers. This contribution may be made at our website or by mail (addresses at the bottom of this page). Any additional tax-deductible contributions to make our ongoing work possible will be accepted gratefully and used well.

Please follow through promptly on your good intentions. Stay legal, and do the right thing.

### SCHOOLS, DISTRICTS, and HOMESCHOOL CO-OPS:

*PURCHASE Option:* Order a book in quantities equal to the number of target classrooms or homes, and receive quantity discounts. If you order 5 books or downloads, for example, then you have unrestricted use of this curriculum for any 5 classrooms or families per year for the life of your institution or co-op.

**2-9 copies of any title:** 90% of current catalog price + shipping.

**10+ copies of any title:** 80% of current catalog price + shipping.

*ROYALTY/LICENSE Option:* Purchase just one book or download *plus* photocopy or printing rights for a designated number of classrooms or families. If you pay for 5 additional Licenses, for example, then you have purchased reproduction rights for an entire book or download edition for any **6** classrooms or families per year for the life of your institution or co-op.

**1-9 Licenses:** 70% of current catalog price per designated classroom or home.

**10+ Licenses:** 60% of current catalog price per designated classroom or home.

### WORKSHOPS and TEACHER TRAINING PROGRAMS:

We are grateful to all of you who spread the word about TOPS. Please limit copies to only those lessons you will be using, and collect all copyrighted materials afterward. No take-home copies, please. Copies of copies are strictly prohibited.

# CONTENTS

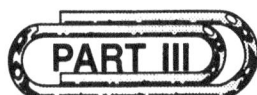

# A TOPS Model for Effective Science Teaching...

***If science were only a set of explanations*** and a collection of facts, you could teach it with blackboard and chalk. You could assign students to read chapters and answer the questions that followed. Good students would take notes, read the text, turn in assignments, then give you all this information back again on a final exam. Science is traditionally taught in this manner. Everybody learns the same body of information at the same time. Class togetherness is preserved.

### But science is more than this.

Science is also process — a dynamic interaction of rational inquiry and creative play. Scientists probe, poke, handle, observe, question, think up theories, test ideas, jump to conclusions, make mistakes, revise, synthesize, communicate, disagree and discover. Students can understand science as process only if they are free to think and act like scientists, in a classroom that recognizes and honors individual differences.

Science is *both* a traditional body of knowledge *and* an individualized process of creative inquiry. Science as process cannot ignore tradition. We stand on the shoulders of those who have gone before. If each generation reinvents the wheel, there is no time to discover the stars. Nor can traditional science continue to evolve and redefine itself without process. Science without this cutting edge of discovery is a static, dead thing.

Here is a teaching model that combines the best of both elements into one integrated whole. It is only a model. Like any scientific theory, it must give way over time to new and better ideas. We challenge you to incorporate this TOPS model into your own teaching practice. Change it and make it better so it works for you.

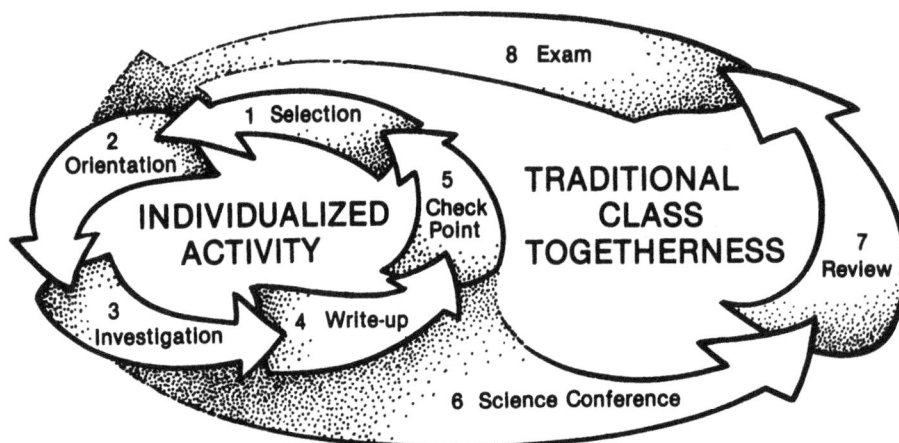

## 1. SELECTION

Doing TOPS is as easy as selecting the first task card and doing what it says, then the second, then the third, and so on. Working at their own pace, students fall into a natural routine that creates stability and order. They still have questions and problems, to be sure, but students know where they are and where they need to go.

Students generally select task cards in sequence because new concepts build on old ones in a specific order. There are, however, exceptions to this rule: students might *skip* a task that is not challenging; *repeat* a task with doubtful results; *add* a task of their own design to answer original "what would happen if" questions.

## 2. ORIENTATION

Many students will simply read a task card and immediately understand what to do. Others will require further verbal interpretation. Identify poor readers in your class. When they ask, "What does this mean?" they may be asking in reality, "Will you please read this card aloud?"

With such a diverse range of talent among students, how can you individualize activity and still hope to finish this module as a cohesive group? It's easy. By the time your most advanced students have completed all the task cards, including the enrichment series at the end, your slower students have at least completed the basic core curriculum. This core provides the common background so necessary for meaningful discussion, review and testing on a class basis.

## 3. INVESTIGATION

Students work through the task cards independently and cooperatively. They follow their own experimental strategies and help each other. You encourage this behavior by helping students only *after* they have tried to help themselves. As a resource person, you work to stay *out* of the center of attention, answering student questions rather than posing teacher questions.

When you need to speak to everyone at once, it is appropriate to interrupt individual task card activity and address the whole class, rather than repeat yourself over and over again. If you plan ahead, you'll find that most interruptions can fit into brief introductory remarks at the beginning of each new period.

## 4. WRITE-UP

Task cards ask students to explain the "how and why" of things. Write-ups are brief and to the point. Students may accelerate their pace through the task cards by writing these reports out of class.

Students may work alone or in cooperative lab groups. But each one must prepare an original write-up. These must be brought to the teacher for approval as soon as they are completed. Avoid dealing with too many write-ups near the end of the module, by enforcing this simple rule: each write-up must be approved *before* continuing on to the next task card.

## 5. CHECK POINT

The student and teacher evaluate each write-up together on a pass/no-pass basis. (Thus no time is wasted haggling over grades.) If the student has made reasonable effort consistent with individual ability, the write-up is checked off on a progress chart and included in the student's personal assignment folder or notebook kept on file in class.

Because the student is present when you evaluate, feedback is immediate and effective. A few seconds of this direct student-teacher interaction is surely more effective than 5 minutes worth of margin notes that students may or may not heed. Remember, you don't have to point out every error. Zero in on particulars. If reasonable effort has not been made, direct students to make specific improvements, and see you again for a follow-up check point.

A responsible lab assistant can double the amount of individual attention each student receives. If he or she is mature and respected by your students, have the assistant check the even-numbered write-ups while you check the odd ones. This will balance the work load and insure that all students receive equal treatment.

## 6. SCIENCE CONFERENCE

After individualized task card activity has ended, this is a time for students to come together, to discuss experimental results, to debate and draw conclusions. Slower students learn about the enrichment activities of faster students. Those who did original investigations, or made unusual discoveries, share this information with their peers, just like scientists at a real conference. This conference is open to films, newspaper articles and community speakers. It is a perfect time to consider the technological and social implications of the topic you are studying.

## 7. READ AND REVIEW

Does your school have an adopted science textbook? Do parts of your science syllabus still need to be covered? Now is the time to integrate other traditional science resources into your overall program. Your students already share a common background of hands-on lab work. With this shared base of experience, they can now read the text with greater understanding, think and problem-solve more successfully, communicate more effectively.

You might spend just a day on this step or an entire week. Finish with a review of key concepts in preparation for the final exam. Test questions in this module provide an excellent basis for discussion and study.

## 8. EXAM

Use any combination of the review/test questions, plus questions of your own, to determine how well students have mastered the concepts they've been learning. Those who finish your exam early might begin work on the first activity in the next new TOPS module.

Now that your class has completed a major TOPS learning cycle, it's time to start fresh with a brand new topic. Those who messed up and got behind don't need to stay there. Everyone begins the new topic on an equal footing. This frequent change of pace encourages your students to work hard, to enjoy what they learn, and thereby grow in scientific literacy.

# GETTING READY

Here is a checklist of things to think about and preparations to make before your first lesson.

## ☐ Decide if this TOPS module is the best one to teach next.

TOPS modules are flexible. They can generally be scheduled in any order to meet your own class needs. Some lessons within certain modules, however, do require basic math skills or a knowledge of fundamental laboratory techniques. Review the task cards in this module now if you are not yet familiar with them. Decide whether you should teach any of these other TOPS modules first: *Measuring Length, Graphing, Metric Measure, Weighing* or *Electricity* (before *Magnetism*). It may be that your students already possess these requisite skills or that you can compensate with extra class discussion or special assistance.

## ☐ Number your task card masters in pencil.

The small number printed in the lower right corner of each task card shows its position within the overall series. If this ordering fits your schedule, copy each number into the blank parentheses directly above it at the top of the card. Be sure to use pencil rather than ink. You may decide to revise, upgrade or rearrange these task cards next time you teach this module. To do this, write your own better ideas on blank 4 x 6 index cards, and renumber them into the task card sequence wherever they fit best. In this manner, your curriculum will adapt and grow as you do.

## ☐Copy your task card masters.

You have our permission to reproduce these task cards, for as long as you teach, with only 1 restriction: please limit the distribution of copies you make to the students you personally teach. Encourage other teachers who want to use this module to purchase their *own* copy. This supports TOPS financially, enabling us to continue publishing new TOPS modules for you. For a full list of task card options, please turn to the first task card masters numbered "cards 1-2."

## ☐Collect needed materials.

Please see the opposite page.

## ☐ Organize a way to track completed assignment.

Keep write-ups on file in class. If you lack a vertical file, a box with a brick will serve. File folders or notebooks both make suitable assignment organizers. Students will feel a sense of accomplishment as they see their file folders grow heavy, or their notebooks fill up, with completed assignments. Easy reference and convenient review are assured, since all papers remain in one place.

Ask students to staple a sheet of numbered graph paper to the inside front cover of their file folder or notebook. Use this paper to track each student's progress through the module. Simply initial the corresponding task card number as students turn in each assignment.

## ☐ Review safety procedures.

Most TOPS experiments are safe even for small children. Certain lessons, however, require heat from a candle flame or Bunsen burner. Others require students to handle sharp objects like scissors, straight pins and razor blades. These task cards should not be attempted by immature students unless they are closely supervised. You might choose instead to turn these experiments into teacher demonstrations.

Unusual hazards are noted in the teaching notes and task cards where appropriate. But the curriculum cannot anticipate irresponsible behavior or negligence. It is ultimately the teacher's responsibility to see that common sense safety rules are followed at all times. Begin with these basic safety rules:

1. Eye Protection: Wear safety goggles when heating liquids or solids to high temperatures.
2. Poisons: Never taste anything unless told to do so.
3. Fire: Keep loose hair or clothing away from flames. Point test tubes which are heating away from your face and your neighbor's.
4. Glass Tubing: Don't force through stoppers. (The teacher should fit glass tubes to stoppers in advance, using a lubricant.)
5. Gas: Light the match first, before turning on the gas.

## ☐Communicate your grading expectations.

Whatever your philosophy of grading, your students need to understand the standards you expect and how they will be assessed. Here is a grading scheme that counts individual effort, attitude and overall achievement. We think these 3 components deserve equal weight:

1. Pace (effort): Tally the number of check points you have initialed on the graph paper attached to each student's file folder or science notebook. Low ability students should be able to keep pace with gifted students, since write-ups are evaluated relative to individual performance standards. Students with absences or those who tend to work at a slow pace may (or may not) choose to overcome this disadvantage by assigning themselves more homework out of class.

2. Participation (attitude): This is a subjective grade assigned to reflect each student's attitude and class behavior. Active participators who work to capacity receive high marks. Inactive onlookers, who waste time in class and copy the results of others, receive low marks.

3. Exam (achievement): Task cards point toward generalizations that provide a base for hypothesizing and predicting. A final test over the entire module determines whether students understand relevant theory and can apply it in a predictive way.

# Gathering Materials

Listed below is everything you'll need to teach this module. You already have many of these items. The rest are available from your supermarket, drugstore and hardware store. Laboratory supplies may be ordered through a science supply catalog.

Keep this classification key in mind as you review what's needed:

| *special in-a-box materials:* | general on-the-shelf materials: |
|---|---|
| Italic type suggests that these materials are unusual. Keep these specialty items in a separate box. After you finish teaching this module, label the box for storage and put it away, ready to use again the next time you teach this module. | Normal type suggests that these materials are common. Keep these basics on shelves or in drawers that are readily accessible to your students. The next TOPS module you teach will likely utilize many of these same materials. |
| (substituted materials): | *optional materials: |
| Parentheses enclosing any item suggests a ready substitute. These alternatives may work just as well as the original, perhaps better. Don't be afraid to improvise, to make do with what you have. | An asterisk sets these items apart. They are nice to have, but you can easily live without them. They are probably not worth an extra trip to the store, unless you are gathering other materials as well. |

Everything is listed in order of first use. Start gathering at the top of this list and work down. Ask students to bring recycled items from home. The teaching notes may occasionally suggest additional student activity under the heading "Extensions." Materials for these optional experiments are listed neither here nor in the teaching notes. Read the extension itself to find out what new materials, if any, are required.

Needed quantities depend on how many students you have, how you organize them into activity groups, and how you teach. Decide which of these 3 estimates best applies to you, then adjust quantities up or down as necessary:

$Q_1 / Q_2 / Q_3$

**Single Student:** Enough for 1 student to do all the experiments.
**Individualized Approach:** Enough for 30 students informally working in 10 lab groups, all self-paced.
**Traditional Approach:** Enough for 30 students organized into 10 lab groups, all doing the same lesson.

| KEY: | *special in-a-box materials* (substituted materials) | | general on-the-shelf materials *optional materials |
|---|---|---|---|

| | | | |
|---|---|---|---|
| 1/10/10 | *pkgs birthday candles, not dripless* | 2/20/20 | medium test tubes |
| .1/1/1 | cup modeling clay | 1/1/1 | source of water and sink or large tub |
| 4/40/40 | tall baby food jars, 6 ounce size | 1/1/1 | *pkg calcium hydroxide – see notes 6* |
| 1/10/10 | pint jars with lids | 1/1/1 | roll plastic wrap |
| 2/20/20 | books matches | 1/2/10 | teaspoons |
| 1/10/10 | household candles — see notes 1 | 1/1/1 | *pkgs active dry yeast – see notes 9* |
| 1/10/10 | tuna fish cans for match disposal | 1/2/2 | *bottles hydrogen peroxide* |
| 1/1/1 | wall clock (watch with a second hand) | 1/10/10 | Popsicle sticks |
| 1/10/10 | empty toilet paper rolls | 1/3/10 | graduated cylinder, 100 mL size |
| 1/10/10 | scissors | 1/1/1 | pkg fine-grade, pure steel wool balls |
| 1/10/10 | plastic produce bags | 1/1/1 | bottle chlorine bleach |
| 1/10/10 | plastic sandwich bags | 2/15/20 | wooden clothespins |
| 2/20/20 | rubber bands | 1/3/10 | *hand calculators |
| 1/10/10 | rubber tubes, at least .5 cm dia and 30 cm long | 1/10/10 | *large test tubes |
| 1/10/10 | rolls masking tape | 1/2/10 | medium-sized nails |
| 1/10/10 | large tubs | 1/3/10 | magnets |
| 1/10/10 | pieces easy-to-bend wire, at least 30 cm long | 1/1/1 | roll aluminum foil |
| 1/10/10 | metric rulers | 1/10/10 | medium-sized washers |
| 1/2/4 | *wire cutters | 1/3/10 | plastic lids from margarine tubs |
| 2/20/20 | size-D batteries, dead or alive | 1/3/10 | quart jars |
| 1/4/10 | tablespoons | 1/1/1 | bottle liquid soap |
| 1/1/1 | bottle vinegar | 1/3/10 | shallow saucers |
| 1/1/1 | box baking soda | 1/1/1 | bottle 70% isopropyl alcohol – see notes 16 |
| 1/2/2 | rolls paper towels | 1/10/10 | eye droppers and * eye-dropper bottles |
| 1/10/10 | *index cards | 1/2/10 | dictionaries |

# Sequencing Task Cards

This logic tree shows how all the task cards in this module tie together. In general, students begin at the bottom of the tree and work up through the related branches. As the diagram suggests, upper level activities build on lower level activities.

At the teacher's discretion, certain activities can be omitted, or sequences changed, to meet specific class needs. The only activities that must be completed in sequence are indicated by leaves that open *vertically* into the ones above them. In these cases the lower activity is a prerequisite to the upper.

When possible, students should complete the task cards in the same sequence as numbered. If time is short, however, or certain students need to catch up, you can use the logic tree to identify concept-related *horizontal* activities. Some of these might be omitted, since they serve only to reinforce learned concepts, rather than introduce new ones.

On the other hand, if students complete all the activities at a certain horizontal concept level, then experience difficulty at the next higher level, you might move back down the logic tree to have students repeat specific key activities for greater reinforcement.

For whatever reason, when you wish to make sequence changes, you'll find this logic tree a valuable reference. Parentheses in the upper right corner of each task card allow you total flexibility; they are left blank so you can pencil in sequence numbers of your own choosing.

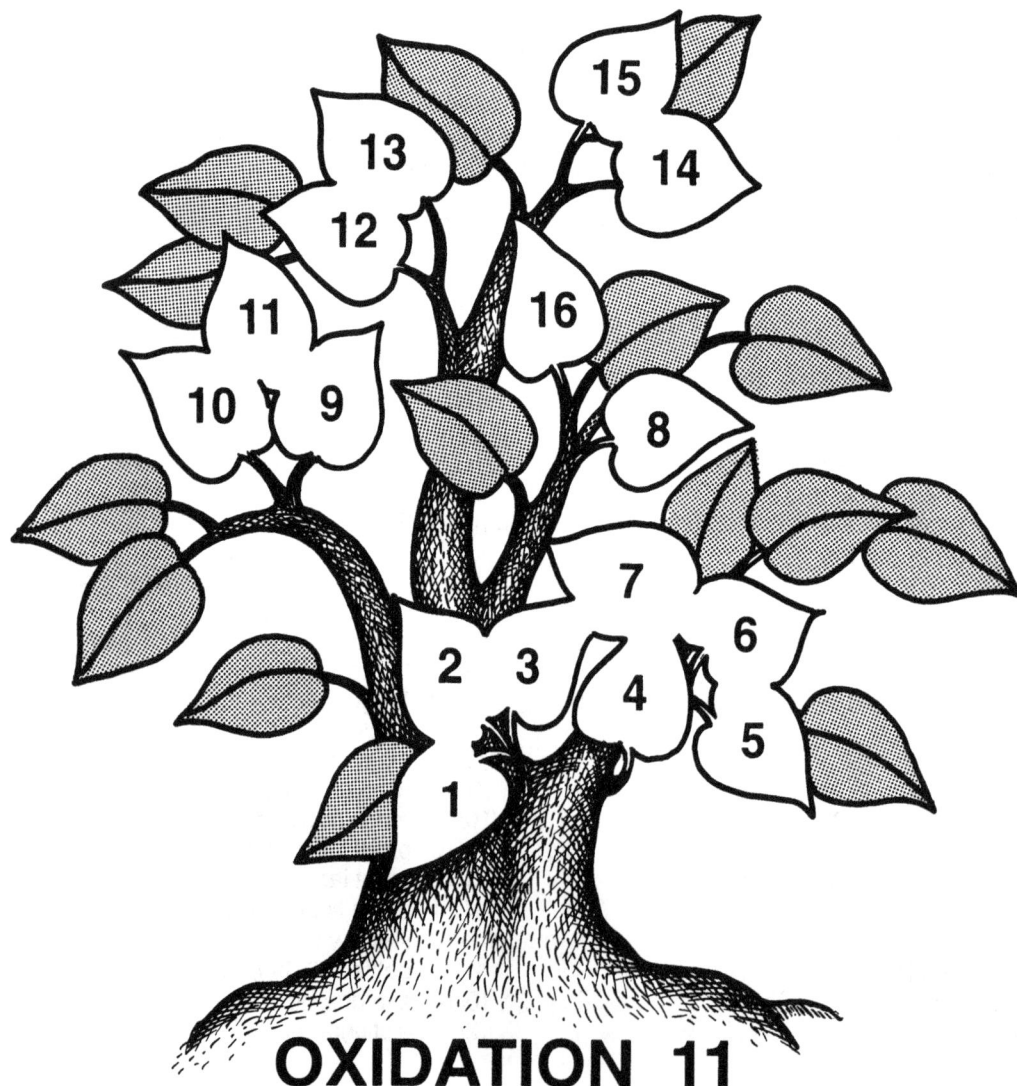

# OXIDATION 11

# LONG-RANGE OBJECTIVES

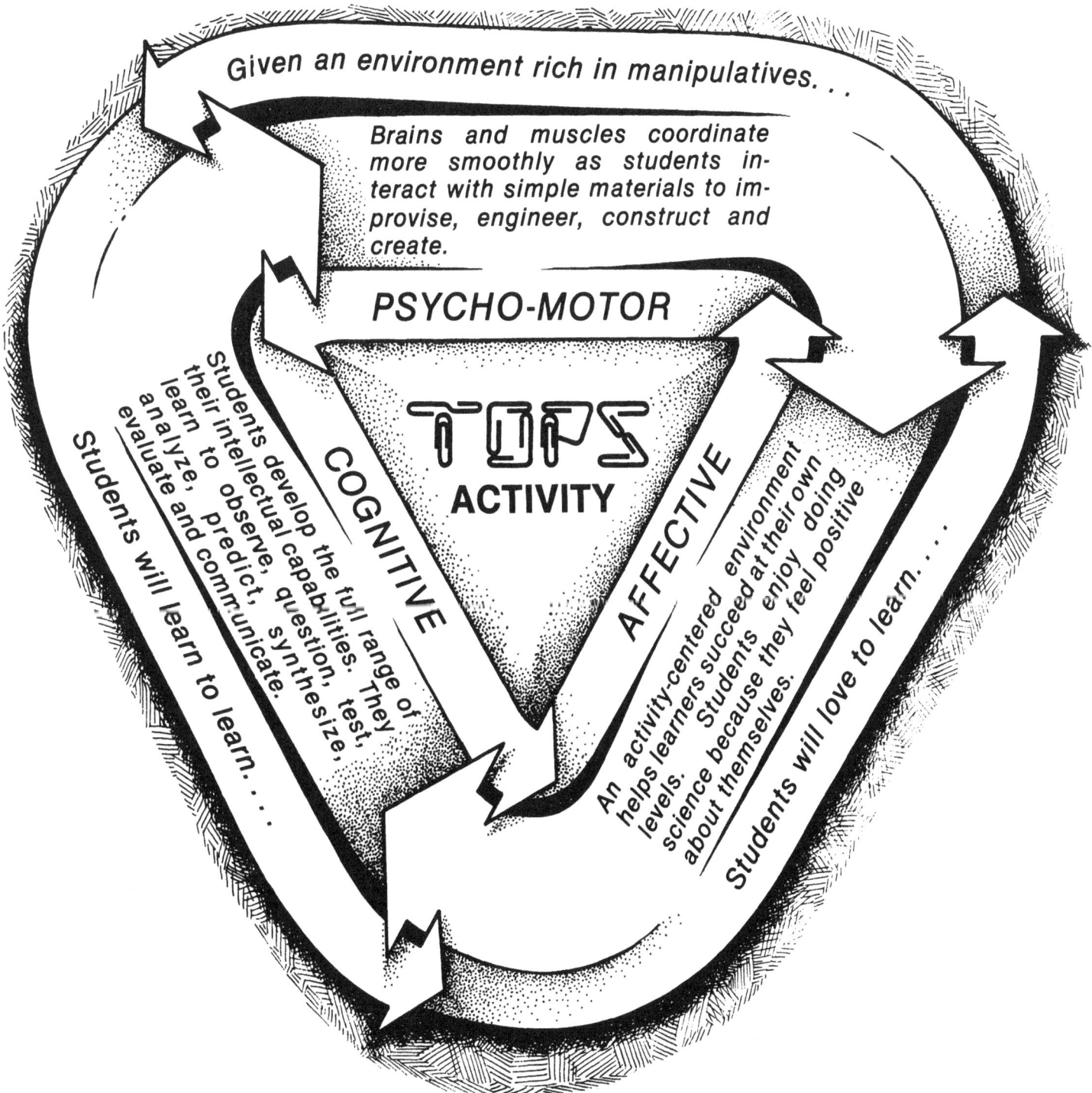

Given an environment rich in manipulatives. . .

Brains and muscles coordinate more smoothly as students interact with simple materials to improvise, engineer, construct and create.

*PSYCHO-MOTOR*

TOPS ACTIVITY

*COGNITIVE*

Students develop the full range of their intellectual capabilities. They learn to observe, question, test, analyze, predict, synthesize, evaluate and communicate.

Students will learn to learn. . . .

*AFFECTIVE*

An activity-centered environment helps learners succeed at their own levels. Students enjoy doing science because they feel positive about themselves.

Students will love to learn. . . .

# Review / Test Questions

Photocopy the questions below. On a sheet of blank paper, cut and paste those questions you want to use in your test. Include questions of your own design, as well. Place all these questions on a single page for students to answer on another paper, or leave space for student responses after each question, as you wish. Duplicate a class set and your custom-made test is ready to use. Use leftover questions as a review in preparation for the final exam.

### tasks 1, 3, 4
Small fires may be put out be covering them with a blanket. Why is this effective?

### tasks 1-3
You are sealed in a small cave by a landslide that has covered your exit route. Rescue workers have brought in heavy equipment to dig you out, but it will probably be many hours before you see the light of day. A candle and matches are in your backpack. You think about lighting the candle to chase away the darkness. Should you?

### tasks 1-3, 9-13
What do a burning candle, a living person and a rusting nail have in common?

### task 3
You wish to transfer some oxygen gas from a large bottle to a small test tube. Diagram how to do this.

### task 4
You are facing an immense prairie fire, with a huge wall of fire rising straight up toward the sky. Which way do you feel the wind blowing? Why?

### tasks 4-5
A candle burns at the bottom of a small jar like this.

a. Draw convection currents associated with the burning candle.
b. You add vinegar and baking soda to the jar without touching the flame or getting it wet. Will the candle go out? Explain.

### task 5
Carbon dioxide is a good fire extinguisher. Explain why.

### tasks 5-12
Three trays are filled with test tubes containing these gases.

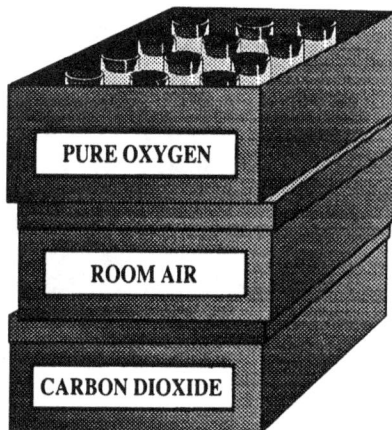

PURE OXYGEN

ROOM AIR

CARBON DIOXIDE

a. Explain how to use a burning splint to verify that each box is correctly labeled.
b. How would you use limewater to identify each gas?
c. Steel wool?

### task 6
You pour carbon dioxide gas into a jar of limewater, cap the top with your hand, and shake it. Why does the jar stick to your hand?

### tasks 6-7
Plants consume carbon dioxide. Design a demonstration to show that this is so. Include both an experiment and a control.

### task 8
a. Write a balanced equation for the combustion of candle wax ($C_{28}H_{58}$).
b. Write a balanced equation for the oxidation of glucose ($C_6H_{12}O_6$).

### tasks 8, 10
Copper (Cu) has a shiny metallic luster, while copper oxide ($Cu_2O$) looks dull brown.
a. How does the appearance of a new penny change with age? Why?
b. Write a balanced equation.

### task 8,16
Write a balanced equation for the combustion of methane gas ($CH_4$).

# Answers

### tasks 1, 3, 4
Putting a blanket over a fire deprives it of the continuous supply of oxygen that it needs to keep burning.

### tasks 1-3
No. There may not be enough oxygen in the small cave to sustain both you and the candle before the rescue workers break through. It is better to sit in the dark and conserve the precious oxygen that remains.

### tasks 1-3, 9-13
A burning candle, a living person, and a rusting nail all consume oxygen (oxidizing wax, food or iron) and produce energy.

### task 3
Transfer the oxygen underwater by bubbling it into an inverted test tube full of water. This will displace the water inside, replacing it with oxygen.

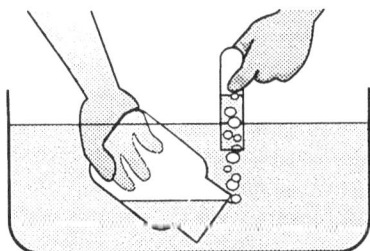

### task 4
You feel the wind at your back because of convection currents. Expanding hot air rises straight up with the flames, while fresh air with oxygen flows in from all sides to feed the flames.

### tasks 4-5
a. Hot air rises up from the center of the jar while cool air is pulled in down the sides of the glass.

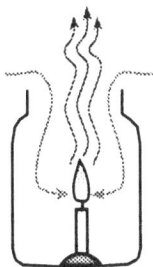

b. Yes, the flame will die. Carbon dioxide and vinegar react to produce carbon dioxide, a heavy gas that does not support combustion. This gas will accumulate in the bottom of the glass and rise until it surrounds the candle flame, displacing the oxygen that sustains it.

### task 5
Carbon dioxide won't burn, and it's heavier than air. It settles over fire, displacing the oxygen that feeds it, and puts it out.

### tasks 5-12
a. Glowing splint: It bursts into flame in a test tube of pure oxygen; it remains glowing and gradually dies in a tube of room air; it extinguishes immediately in a tube of carbon dioxide.

b. Limewater: Limewater does not react when poured into a test tube of oxygen; it forms a surface crust of calcium carbonate in a test tube of room air; it forms a milky white precipitate of calcium carbonate in a test tube of carbon dioxide.

c. Steel wool: Invert each test tube in water and push a ball of steel wool up inside. The steel wool rusts heavily in oxygen and water rises high into the tube; it rusts moderately in room air and water rises 1/5 of the way up the tube; it doesn't rust in pure carbon dioxide, so water won't rise in the tube at all.

### task 6
Limewater reacts with carbon dioxide gas to form a solid calcium carbonate precipitate. This consumes the gas, creating a partial vacuum inside the jar that draws in the palm of your hand.

### tasks 6-7
Mix baking soda and vinegar in a large container. Pour the heavy gas into 2 quart jars. Invert one over a small potted plant to serve as your experiment. Invert the other over a pot of equal size that contains everything except a plant. This is your control. After a day or so, bubble gas from each jar into 2 test tubes using underwater displacement. Add limewater to the gas inside each test tube, and shake it to mix. Higher levels of unused carbon dioxide in the control air should produce more milky white calcium carbonate precipitate than in the experiment containing the plant air.

### task 8
a. $2 C_{28} H_{58} + 85 O_2 \longrightarrow$
$56 CO_2 + 58 H_2O + Energy$

b. $C_6H_{12}O_6 + 6 O_2 \longrightarrow$
$6 CO_2 + 6 H_2O + Energy$

### tasks 8, 10
a. Over time, oxygen in the air combines with copper in the penny to oxidize its shiny surface to a dull brown.

b. $4 Cu + O_2 \longrightarrow 2 Cu_2O$

### tasks 8, 16
$CH_4 + 2 O_2 \longrightarrow CO_2 + 2 H_2O$

# Review / Test Questions (continued)

## task 9
Explain how to use a glowing splint to decide whether a jar of unknown gas contains abundant oxygen.

## tasks 10-11 A
Describe what happens when a nail rusts.

## tasks 10-11 B
Suppose you put wet steel wool in each test tube, fill them with oxygen, room air, or carbon dioxide as shown, then seal airtight with balloon membranes and rubber bands. Redraw this diagram showing how each test tube might look after several weeks. Give reasons to support each prediction.

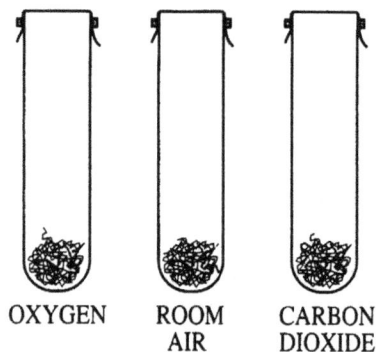

OXYGEN    ROOM AIR    CARBON DIOXIDE

## tasks 10-11 C
A company sells cylinders of gas, claiming that they contain 50% oxygen. How might you test this claim by experiment?

## tasks 10, 11, 14, 15
Three pieces of steel wool were dipped in oil, and another three pieces were left untreated. They were placed in test tubes (in various positions) and inverted over water. After 24 hours the water levels rose only in the test tubes that contained untreated steel wool.

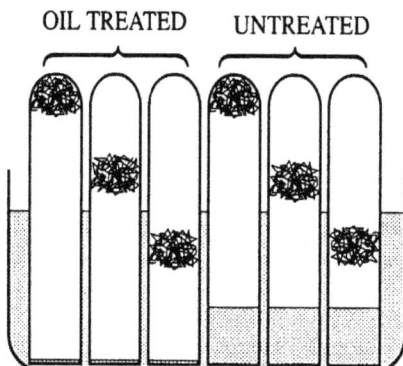

OIL TREATED     UNTREATED

a. Why did water rise in some of the tubes, but not others?
b. Identify an important variable in this experiment that affected the results.
c. Identify an unimportant variable in this experiment that did not affect the results.
d. All important variables in this experiment are controlled. How do you know?

## task 12
You are assigned the task of burning a nail. Explain how to do it.

## tasks 12-13
Is iron oxide magnetic?

## tasks 14-15 A
A glass tube has one end sealed with a balloon membrane. The open end is lowered over a burning candle and stuck into a "pancake" of clay, forming an airtight seal.

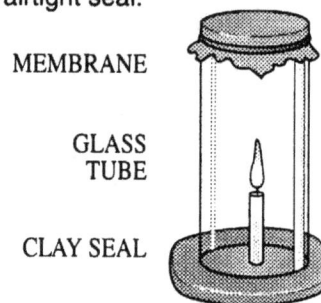

MEMBRANE

GLASS TUBE

CLAY SEAL

a. Will the membrane bulge out? Why?
b. Will the membrane be drawn in? Why?

## tasks 14-15 B
A good scientists is always ready to replace a good hypothesis with a better one. Why not just start out with the best hypothesis and be done with it?

## task 15
Burning candles are placed under different sized containers as illustrated.

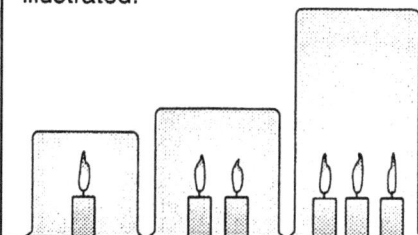

a. Identify 2 important variables in this experiment.
b. Why is it difficult to predict which candles will go out first?

## task 16
Design a demonstration to show that a car engine produces carbon dioxide and water vapor when it burns gasoline.

# Answers (continued)

## task 9

Introduce the glowing splint into the jar. If the splint reignites, the jar is rich in oxygen. If the splint continues to glow and gradually goes out, the jar has an oxygen concentration similar to room air. If the splint stops glowing immediately, the jar contains little or no oxygen.

## tasks 10-11 A

Iron in the nail is slowly oxidized to (a hydrated form of) iron oxide.

## tasks 10-11 B

Oxygen gas reacts with iron to produce various forms of solid iron oxide (rust). This removes…

…all the gas from the tube that contains pure oxygen (except for water vapor).

OXYGEN

…one fifth of the gas from the test tube that contains room air, since it is comprised of that proportion of oxygen.

ROOM AIR

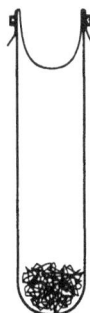

…no gas from the tube that contains carbon dioxide, since this gas does not react with iron (under these conditions).

CARBON DIOXIDE

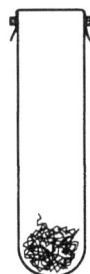

## tasks 10-11 C

Put balls of steel wool in several test tubes. Fill them with water and invert them in beakers of water. Then bubble the gas you wish to test into each tube, displacing all the water inside. After several days, the rusting balls of steel wool will consume all of the oxygen in the gas. If its purity is 50%, water will consistently rise halfway in each tube.

## tasks 10, 11, 14, 15

a. The untreated steel wool balls rusted. Water rose in these test tubes to replace oxygen gas that combined with the steel wool to form solid iron oxide. The oil-treated steel wool balls didn't rust, so no oxygen was consumed.
b. Important variable: Some steel wool balls were dipped in oil and some were not.
c. Unimportant variable: The steel wool balls were placed in different positions within each tube.
d. All important variables are controlled, because the results are reproducible.

## task 12

File it to a pile of iron shavings, then ignite these tiny pieces in an atmosphere of abundant oxygen.

## tasks 12-13

Yes and no. Iron oxide comes in 3 forms. $Fe_3O_4$ is magnetic, while $Fe_2O_3$ and $FeO$ are not magnetic.

## tasks 14-15 A

a. Yes. The balloon membrane will bulge out because the burning candle heats air trapped inside the tube, causing it to expand.
b. Yes. When the flame uses up available oxygen and goes out, air in the tube will cool and contract, drawing the balloon membrane back in. It will be drawn below its starting position, because part of the heated air originally inside the tube expanded away before the mouth of tube was sealed in clay.

## tasks 14-15 B

A hypothesis is an educated guess based on what is known. New experiments and new discoveries are always revealing new evidence, requiring us to continually refine our ideas.

## task 15

a. The 2 most important variables are the size of each jar and the number of candles in each jar.
b. It is difficult to predict which candles will go out first, because these 2 variables work against each other: the large jar has more oxygen, but 3 candles to consume it more rapidly; the small jar has less oxygen, but only 1 candle to consume it more slowly.

## task 16

Hold a clean, cool jar over the tail pipe of a car while its engine is running. Water vapor will immediately condense inside the jar. Pour limewater into this same jar and it will turn milky white, indicating the presence of carbon dioxide.

# TEACHING NOTES
## For Activities 1-16

**Task Objective (TO)** discover that candles require a constant supply of oxygen to burn. To understand this process as an energy-producing oxidation reaction.

---

## CANDLE COMBUSTION  ○                    Oxidation (   )

1. Stand 2 birthday candles in small lumps of clay. Put one next to a small jar, the other next to a large jar.

*READY?*

2. Light the candles, then set a jar over each one at the same time. Write your observations.

3. Blow fresh air into each jar, and repeat the experiment. Time how long each candle burns inside each jar.

4. *Combustion* (burning) is a process that uses up both fuel and oxygen.
   a. Compare a new candle with a used one. What fuel is being consumed?
   b. Why do burning times vary with jar size?

5. When a fuel burns, it *oxidizes* (combines with oxygen).
   a. What gets oxidized in this experiment?
   b. Does this oxidation reaction absorb energy or release energy? Explain.

© 1995 by TOPS Learning Systems                                                                1

---

## Introduction

Light a candle. Observe how a piece of paper (like hair and clothing) can catch fire at a distance. Discuss the safe and responsible use of open flames, disposal of hot matches, etc.

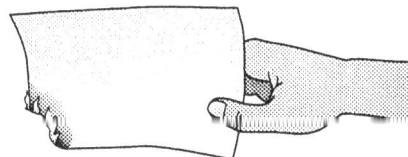

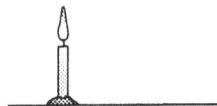

## Answers / Notes

2. Both covered candles burn for a little while, then go out. The candle covered by the large jar burns longer than the candle covered by the small jar.

3. *Expect variations up to several seconds depending on the size of the candle flame and how thoroughly used air is purged from each container. Here is one result:*
   6 oz. baby food jar: candle burned 7 seconds
   pint jar: candle burned 16 seconds

4a. A used candle is shorter than a new one. This implies that candle wax is the fuel that gets consumed.

4b. The candle burns for a longer time in the larger jar because it contains more oxygen than the smaller jar. (*Air is a mixture of gases that include <u>oxygen</u>. For now, these words can be used interchangeably. Their distinction will emerge naturally over time.*)

5a. Candle wax gets oxidized as it burns.

5b. This oxidation reaction releases energy in the form of heat and light.

## Materials

☐ Birthday candles. Avoid the dripless variety with hollow centers.
☐ Modeling clay.
☐ A small jar. We recommend using a tall 170 g (6 ounce) baby food jar. Smaller sizes may not provide enough "head room" for the burning candle.
☐ A large jar, pint or quart size.
☐ Matches.
☐ A household candle (optional). Keeping 1 continuously lighted pilot candle in an accessible place will dramatically reduced match consumption.
☐ A tuna fish can or other shallow container for the safe disposal of used matches.
☐ A wall clock or a watch with a second hand.

**(TO)** compare human use of oxygen to that of a burning candle.

---

## HUMAN RESPIRATION ◯      Oxidation ( )

1. Cut an empty toilet paper roll into 2 equal tubes. Rubberband the mouth of a sandwich bag around one, a plastic produce bag around the other.

SANDWICH BAG

PRODUCE BAG

2. Take a deep breath and hold your nose. Breathe in and out through your mouth, as normally as possible, into the larger bag.
    a. Look at a clock. How long can you use the same air over and over before you feel uncomfortable?
    b. How did your breathing change over time?
    c. Did you collect anything in the bag besides "used" air?

3. Now exhale fully into the small bag, allowing excess air to leak past the mouthpiece. Once again, hold your nose and breathe through your mouth as normally as possible.
    a. How long did you rebreathe the same air? Compare this result with step 2a.
    b. Compare human *respiration* (breathing) to candle combustion.

4. A candle oxidizes wax.
    a. What do you think *you* oxidize?
    b. Does the oxidation reaction in your body absorb or release energy? Explain.

© 1995 by TOPS Learning Systems

2

---

## Answers / Notes

2a. Two minutes, more or less. *(Answers will vary widely depending on physiology and individual comfort levels. Competitive students will naturally try to outdo each other. The Guiness Book of Records lists the world record for breath-holding at well over 13 minutes!)*

2b. Over time, both the rate of breathing and the amount of air inhaled in each breath increased dramatically.

2c. Yes. A clear liquid has condensed in the bag, presumably water.

3a. One minute, more or less. *(This result should be significantly less than when breathing into the larger produce bag.)*

3b. Human respiration and candle combustion both consume oxygen. The larger this oxygen supply, the longer both people and candles can survive before they "burn out."

4a. We oxidize the food we eat. *(We convert this into blood sugars for immediate oxidation, into stored body fats for later oxidation.)*

4b. Oxidized food releases energy. We are warm bodies that do work.

## Materials

☐ An empty toilet paper roll.
☐ Scissors.
☐ A plastic produce bag.
☐ A plastic sandwich bag.
☐ Two rubber bands.
☐ A wall clock or a watch with a second hand.

**(TO)** collect gas by displacing water. To show that exhaled air contains less oxygen than inhaled air.

---

## I FEEL FAINT                                    O          Oxidation (     )

1. Wrap masking tape around the end of a rubber tube to make a sanitary mouthpiece. You will blow air through it.

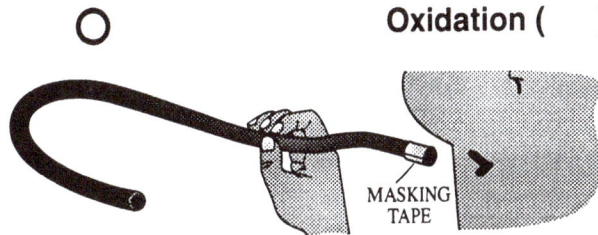

MASKING TAPE

2. Devise a way to collect and seal only your breath (and no other air) inside a pint jar. Hint: use these materials.

   a. Detail your method in words and pictures.

   b. Hold your breath as long as you can, then collect it inside a pint jar *before* you draw a breath of fresh air. Close with a lid.

TUB OF WATER

BIG BREATH (Hold as long as possible)

TUBE

PINT JAR WITH LID

3. Show that this pint of "breathed" air contains little or no oxygen. Explain your method and results.

© 1995 by TOPS Learning Systems                                                      3

---

## Answers / Notes

2, 2a. *Don't give away the answer. Allow your students the satisfaction of experimenting by trial and error until they make their own discoveries.*

First fill the pint jar completely with water, then invert it in the large tub. *(If the depth of the water is less than the height of the pint jar, first fill the jar brimful and seal the water inside with a lid. Then turn the jar over and remove the lid underwater.)*

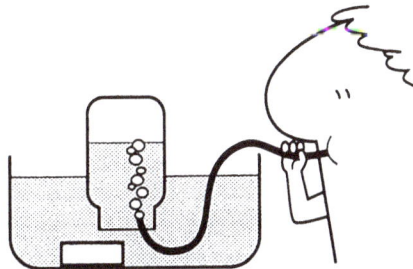

Then blow air through the rubber tube so it bubbles up inside the jar, displacing the water inside. When the jar is full of your breath, seal it underwater with the lid before you take it out of the water.

3. Invert the jar so all residual water collects in the lid. Then unscrew it and quickly place the jar over a burning candle. The flame will extinguish instantly if the breath was held as long as possible before collection. Students should then compare this result with how long the candle burned under a pint jar of room air.

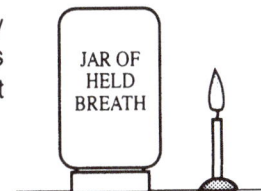

JAR OF HELD BREATH

## Materials

☐ Rubber or plastic tubing with a minimum length of 30 cm (1 foot).
☐ Masking tape.
☐ A tub of water. If the depth of the tub is less than the height of the pint jar, include an extra container for pouring water.
☐ A pint jar with lid.
☐ A birthday candle with clay base.
☐ Matches (a pilot candle).
☐ A disposal can for matches.
☐ A wall clock or watch with a second hand.

**(TO)** study how convection currents sustain a candle that burns in the bottom of a jar. To recognize that water vapor is a product of both combustion and respiration.

---

## BOTTOM BURNER                    ○                    Oxidation (   )

1. Loop the end of a 30 cm piece of wire around a pencil. Press the clay base of a birthday candle into this loop.

2. Shape this wire so the candle stands centered at the bottom of a small jar, while the other end of the wire is looped and rubber-banded to a battery "handle."

3. Use the handle to lift the candle from the jar, light it, and lower it back inside the uncovered jar. Does it keep burning?

4. Hold another small jar, mouth to mouth, over the first.
   a. What happens inside the top jar?
   b. What happens inside the bottom jar?

5. Draw a side view of the candle and jar, showing how *convection currents* (air streams driven by heat) supply a steady flow of oxygen to the candle flame. Label and explain your drawing.

WIRE LOOP

6. Blow your breath into a cool, dry jar and observe closely. Name 2 ways you are like a burning candle. (Save your candle holder to use again.)

4

---

### Answers / Notes

3. Yes, the candle remains burning in the bottom of the jar.

4a. The inside of the top jar turns foggy with condensed water vapor. *(The inside of the top jar must be initially clean and dry to best see this effect. Cooling the jar under cold tap water or in a refrigerator will increase the amount of condensation.)*

4b. The candle flame goes out in the bottom jar. Then both jars fill with smoke.

5.
UPWARD ⬆ CONVECTION

SIDE-STREAMS SUPPLY OXYGEN

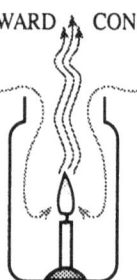

The candle flame heats the air directly above it, causing it to expand and rise. This upward convection current pulls cool fresh air in from the sides to resupply the candle with fresh oxygen.

6. Breathing into the cool, dry jar clouds it with water vapor. Both the candle and I consume oxygen, and both produce water vapor.

### Demonstration

Trace a path of fresh air flowing into a baby food jar with a punk, a smoldering Popsicle stick, or other source of smoke. Hold it just outside the rim of the jar. Some of the rising smoke will be drawn over the lip, and down inside the jar.

### Materials

☐ A foot long (30 cm) piece of easy-to-bend wire.
☐ A metric ruler (optional).
☐ Wire cutters (optional).
☐ A birthday candle stuck in a small clay base. A short, used candle burning low in the jar looks most dramatic.
☐ A size-D battery, dead or alive.
☐ A rubber band.
☐ Two tall baby food jars, clean and dry.
☐ Matches (a pilot candle).
☐ A disposal can for matches.

**(TO)** produce carbon dioxide gas and pour it over a candle flame. To design a fire extinguisher.

---

## IT'S A GAS       O       Oxidation ( )

1. Add 3 tablespoons (36 mL) vinegar to a pint jar. Dry the spoon, then add 1 rounded tablespoon (18 mL) of baking soda. Cover the mouth of the jar with an index card.

2. What evidence do you see that this chemical reaction is taking place: **Vinegar + Baking Soda ⟶ Carbon Dioxide Gas**

3. Use your candle holder and battery to hold a short burning candle in the bottom of a small jar, as before.

4. Pour carbon dioxide gas (not the liquid underneath) over your burning candle. Explain your observations in terms of one gas *displacing* (taking the place of) another.

5. Add a pinch of baking powder to half a test tube of vinegar. Cap it tightly with your thumb and invert over a sink.
   a. What happens? Why?
   b. Design a vinegar and baking soda fire extinguisher. Include operating instructions with your design.

© 1995 by TOPS Learning Systems       5

---

## Introduction

Review, if necessary, the notion of *displacement*. Pour half a test tube of water (tinted with food coloring for visibility) into a half test tube of corn oil or other immiscible liquid.
a. Describe what happened. (Water displaced corn oil.)
b. Will corn oil displace water if you reverse the procedure? (No. Corn oil is lighter than water.)

## Answers / Notes

2. The mixture of vinegar and baking soda produces numerous gas bubbles, possibly carbon dioxide, that rise in the jar like foam.

4. The candle goes out. Heavy carbon dioxide gas pours out of the pint jar and fills the smaller jar, displacing lighter oxygen gas that formerly surrounded and sustained the candle flame.

5a. Vinegar reacts with baking soda to produce carbon dioxide gas. This builds pressure on the liquid in the inverted test tube until it squirts out from under your thumb.

5b. *If time and materials are available, encourage students to design a fire extinguisher from available lab materials they can actually assemble. The fire extinguisher should keep the baking powder separated from the vinegar until it is ready to use. Here is one example:*

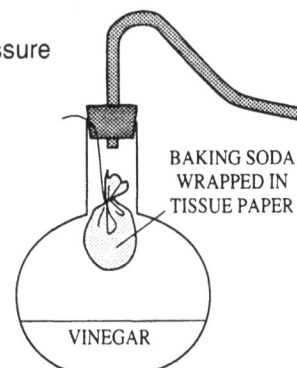

BAKING SODA WRAPPED IN TISSUE PAPER

VINEGAR

OPERATING INSTRUCTIONS:
Turn upside down. Aim the stream of vinegar and carbon dioxide foam at the fire.

## Materials

- A pint jar.
- A tablespoon
- Vinegar.
- Baking soda.
- A paper towel.
- An index card. You may substitute a canning lid, but no screw-on ring!
- A baby food jar.
- The candle holder, including candle and battery, from the last activity. Short candles work best.
- Matches (a pilot candle).
- A disposal can for matches.
- A test tube.
- A sink (large tub).

**(TO)** learn how to test for the presence of carbon dioxide gas with limewater. To set up a control as a standard of comparison.

---

## LIMEWATER REACTION  ◯                    Oxidation (    )

1. Fill 2 small jars about 1/4 full with limewater. Label one C for <u>C</u>ontrol, the other E for <u>E</u>xperiment.

2. Rubberband the mouth of your Control with a small square of plastic wrap. Reserve another rubber band and square of wrap for step 3.

3. Make a pint of carbon dioxide as before. Pour all this gas (not the liquid underneath) into your Experiment, then seal it as quickly as possible.

4. Describe what happens (if anything) in each jar.

5. Interpret your observations in terms of this equation:

### Limewater + Carbon Dioxide Gas ⟶ Calcium Carbonate

CARBON DIOXIDE

LIME-WATER

PRECIPITATE (a white solid)

6. Your experiment and control are alike in every respect but one.
   a. What is the *controlled variable* (the one difference)?
   b. Why have a control?

7. Remove the plastic wrap from both jars and set them aside overnight. How does the limewater surface change over time? What can you conclude?

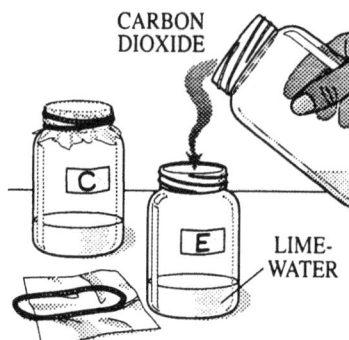

6

---

## Answers / Notes

4. Experiment Jar: The surface of the limewater turns a milky white color which spreads down through the liquid and gradually settles to the bottom of the jar. Plastic wrap at the top of the jar pushes inward.

   Control Jar: There are no noticeable changes. *(Crusty white patches may be floating on top of the limewater, but these were present from the start.)*

5. The milky white color is caused by solid calcium carbonate precipitated from solution. Over time this precipitate settles to the bottom of the jar. Carbon dioxide gas is used up, creating a partial vacuum above the limewater. Higher atmospheric pressure outside the jar pushes the plastic wrap inward. This reaction is not observed in the control jar because it contains room air.

6a. The controlled variable is the different gas inside each jar. The experiment jar holds carbon dioxide gas, while the control jar holds room air.

6b. The control insures that carbon dioxide, not room air or some other variable, reacts with limewater.

7. Overnight, a crusty surface film forms on top of the limewater in both jars. This crust is probably calcium carbonate, formed as carbon dioxide in room air makes contact with the surface of the limewater. *(This crust is thicker in the control jar than in the experiment jar, because there is more calcium hydroxide in its limewater still available to react with carbon dioxide.)*

## Materials

☐ Two baby food jars.

☐ Limewater. Prepare this from hydrated lime (also called calcium hydrate, calcium hydroxide, and garden lime). Purchase in as small a quantity as possible from garden supply or farm stores. Lab grade purity is not required. The solubility of this powder in water is very low. Stir in a level teaspoon (4 mL) into a quart (liter) jar of water, then close with a lid and allow the chalky white mixture to settle about 24 hours. (Limewater left exposed to carbon dioxide in air forms a heavy surface film of calcium carbonate.) Pour off the clear liquid into a second storage jar with lid and label it LIMEWATER $(CaOH_2)$. Add more water to the sediment in your original jar, and set aside to resupply your labeled jar as needed. The shelf life of lime may degrade over time. Check that your limewater still gives a strong positive test for carbon dioxide: fill a baby food jar about 1/6 full and blow bubbles into it through a straw. Within 15 seconds or so, the clear solution should turn the milky white color of precipitated calcium carbonate:

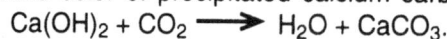

$$Ca(OH)_2 + CO_2 \longrightarrow H_2O + CaCO_3.$$

☐ Masking tape.

☐ Plastic wrap and scissors.

☐ Two rubber bands.

☐ A pint jar.

☐ A tablespoon and teaspoon.

☐ Vinegar and baking soda.

**(TO)** use the limewater test to identify carbon dioxide as a product of both respiration and combustion.

---

## THREE LITTLE JARS          O                    Oxidation (   )

1. Rinse 3 small jars of equal size with water. Cover each jar with a labeled lid identifying the air you put inside.

### a. ROOM AIR:

It's already in the jar.

### b. RESPIRATION AIR:

Hold your breath as long as you can. Collect it by water displacement.

### c. COMBUSTION AIR:

Light a candle in its holder and set it in the bottom of another jar. Hold your test jar inverted over the top until the flame goes out.

2. These jars are all the same except for one controlled variable. What is it?

3. Test the gas in each jar for carbon dioxide: open each lid just wide enough to pour in limewater, close it quickly, and shake the jar.
   a. Report your results.          b. What can you conclude?

© 1995 by TOPS Learning Systems                                    7

---

## Answers / Notes

1b. *Refer students to activity 3 if they forget how to collect gas by water displacement. Though it is not technically necessary to do so, you might suggest that students flush room air out of the hose by blowing into it before collecting gas bubbles in the inverted jar.*

2. The variable that is controlled is the kind of gas in each jar.

3a. The limewater remains clear in the jar of room air, but turns milky white in the respiration air and combustion air.

3b. Carbon dioxide reacts with limewater to produce calcium carbonate, a milky white precipitate. This reaction occurred in two out of three jars, indicating a strong presence of carbon dioxide in air that was breathed and air that supported a burning candle flame. Ordinary room air contained insufficient carbon dioxide to produce a noticeable reaction.

## Materials
- ☐ A source of water.
- ☐ Four 6 ounce baby food jars with 3 lids.
- ☐ Masking tape
- ☐ A tub of water.
- ☐ The rubber tube from activity 3.
- ☐ The candle with holder and battery base from activity 4.
- ☐ Matches.
- ☐ A disposal can for matches.
- ☐ Limewater.

## CHEMICAL OVERVIEW     O     Oxidation (   )

1. Review the (**activity**/step) under each blank, then write the correct word.

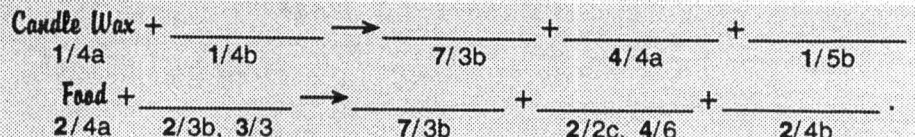

Candle Wax + _____ ⟶ _____ + _____ + _____ .
1/4a    1/4b     7/3b     4/4a     1/5b

Food + _____ ⟶ _____ + _____ + _____ .
2/4a   2/3b, 3/3    7/3b    2/2c, 4/6    2/4b

2. Pair each substance with its correct molecular formula:

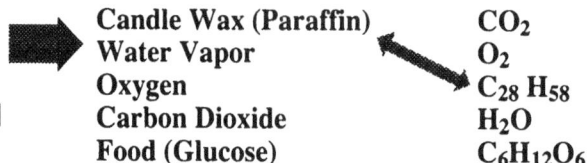

3. Rewrite both equations in step 1 using molecular formulas.

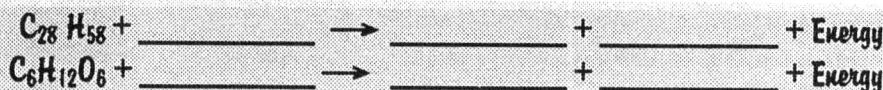

| Candle Wax (Paraffin) | $CO_2$ |
| Water Vapor | $O_2$ |
| Oxygen | $C_{28}H_{58}$ |
| Carbon Dioxide | $H_2O$ |
| Food (Glucose) | $C_6H_{12}O_6$ |

$C_{28}H_{58}$ + _____ ⟶ _____ + _____ + Energy

$C_6H_{12}O_6$ + _____ ⟶ _____ + _____ + Energy

4. Rewrite both equations in step 3 so they balance. Each side must have the same number of <u>C</u>arbon atoms, <u>H</u>ydrogen atoms, and <u>O</u>xygen atoms.

YOU **CAN** CHANGE THE NUMBER OF MOLECULES...

$2\ C_{28}H_{58} + 85\ O_2$

...BUT **NEVER** THE NUMBER OF ATOMS IN A MOLECULE.

© 1995 by TOPS Learning Systems

8

## Introduction

Write this equation on your blackboard. Use it to discuss these ideas about balancing chemical equations:

$$H_2O + energy \longrightarrow H_2 + O_2$$

● It says that when you apply energy (electricity) to a molecule of water, it splits into hydrogen gas and oxygen gas ( a process called electrolysis).
● The horizontal arrow means "equals," "yields," or "gives."
● Each molecule in the equation has a coefficient. If no number is written, the coefficient is understood to be "1."

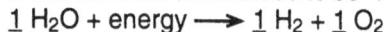

$$\underline{1}\ H_2O + energy \longrightarrow \underline{1}\ H_2 + \underline{1}\ O_2$$

● An equation is balanced when all atoms on both sides of the arrow are equal. Is our equation balanced? No, the left side has 1 oxygen atom, while the right side has 2.
● Balance equations by trial and error, as if you were solving a puzzle. Write different coefficients in front of each molecule until you hit the right combination. Let's write "2" in front of the water molecule to balance oxygen on both sides.

$$\underline{2}\ H_2O + energy \longrightarrow H_2 + O_2$$

This does balance oxygen, but now hydrogen is unequal. Let's write "2" in front of the hydrogen molecule.

$$2\ H_2O + energy \longrightarrow 2\ H_2 + O_2$$

Now there are 4 hydrogen on both sides and 2 oxygen. The equation is balanced.
● When balancing equations, you can always change the coefficients (numbers of molecules), but *never* the numbers of atoms that make up each molecule. Consider:

$$H_2O + energy \longrightarrow H_2 + O$$

This equation is balanced, but incorrectly represents the stable diatomic character of oxygen gas. Single atoms of oxygen are too reactive to remain single for more than an instant.

## Answers / Notes

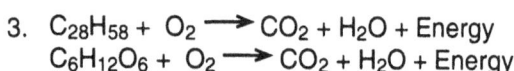

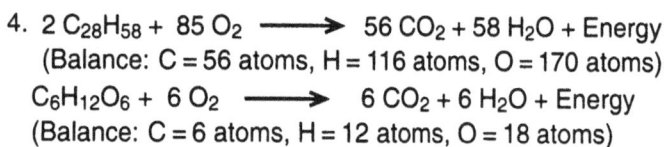

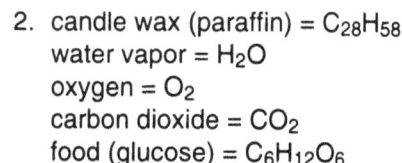

1. Candle Wax + Oxygen ⟶ Carbon Dioxide + Water Vapor + Energy
   Food + Oxygen ⟶ Carbon Dioxide + Water Vapor + Energy

2. candle wax (paraffin) = $C_{28}H_{58}$
   water vapor = $H_2O$
   oxygen = $O_2$
   carbon dioxide = $CO_2$
   food (glucose) = $C_6H_{12}O_6$

3. $C_{28}H_{58} + O_2 \longrightarrow CO_2 + H_2O$ + Energy
   $C_6H_{12}O_6 + O_2 \longrightarrow CO_2 + H_2O$ + Energy

4. $2\ C_{28}H_{58} + 85\ O_2 \longrightarrow 56\ CO_2 + 58\ H_2O$ + Energy
   (Balance: C = 56 atoms, H = 116 atoms, O = 170 atoms)
   $C_6H_{12}O_6 + 6\ O_2 \longrightarrow 6\ CO_2 + 6\ H_2O$ + Energy
   (Balance: C = 6 atoms, H = 12 atoms, O = 18 atoms)

## Materials

☐ Completed assignment sheets from activities 1-4 and 7 to use as reference.

**(TO)** produce oxygen by the catalytic decomposition of hydrogen peroxide. To test for the presence of oxygen-rich gas with a glowing splint.

---

## GLOWING SPLINT                    O                    Oxidation (    )

1. A molecule of hydrogen peroxide is like water, but with 1 extra oxygen atom. Write the formula for both substances.

2. Hydrogen peroxide decomposes (breaks down) to form water and oxygen gas. Write a *balanced* equation (same number of atoms on both sides of arrow).

3. Put a *small* pinch of yeast in the bottom of a clean, dry test tube.
   a. Fill the test tube ¼ full with <u>hydrogen peroxide</u>. Rest it in a small jar.
   b. Describe and interpret the reaction (if any).
   c. Burn a Popsicle stick for about 10 seconds. Blow out the flame, then poke the glowing splint halfway into the test tube. What happens?

4. Set up a control by repeating all of step 3, but substitute <u>water</u> for <u>hydrogen peroxide</u> in step 3a. This is your controlled variable.
   a. What happens differently?
   b. Does the glowing splint reveal that the control test tube contained at least *some* oxygen? Explain.

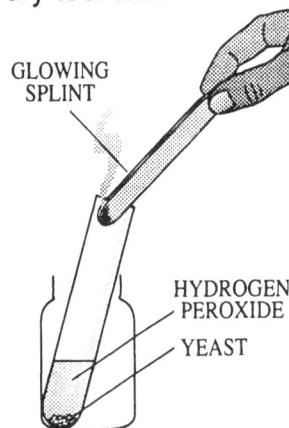

GLOWING SPLINT

HYDROGEN PEROXIDE

YEAST

9

---

### Answers / Notes

1.  water = $H_2O$;  hydrogen peroxide = $H_2O_2$
2.  $2\ H_2O_2 \rightarrow 2\ H_2O + O_2$.

3. *No more than a pinch of yeast is required. As a catalyst, it does not get used up in the reaction. A dry test tube is specified so the yeast granules won't stick to the wet sides of the tube.*

3b. The hydrogen peroxide fizzes on contact with the yeast, producing oxygen gas bubbles as it decomposes. *(Good observers may notice that the bottom of the test tube is slightly warmer than the top, indicating that heat energy is produced.)*

3c. The glowing splint bursts into flame, burns for a few moments, then goes out. *(If too much yeast was used, bubbles may rise in the tube and even overflow the top. These will not extinguish the glowing splint. It reignites in the oxygen-rich gas, just the same.)*

*The splint should be ground dead out in a shallow can or lid and not discarded in a waste paper basket! Close supervision is a must.*

4a. No reaction occurs when water is substituted for hydrogen peroxide as the controlled variable. A glowing splint does not reignite inside the control test tube.

4b. Yes. The splint glow was sustained by oxygen in the control test tube for a few seconds before its dull red glow faded to white ash.

### Materials

☐ Two medium-to-large test tubes of equal size.
☐ Paper towels.
☐ Active dry yeast to use as a catalyst. Find this in the baking section of your grocery store. Potassium iodide (KI) and manganese dioxide ($MnO_2$) also work. $MnO_2$, the traditional catalyst in this reaction, is the black material found in ordinary carbon-zinc dry cells. It is messy to retrieve and use.
☐ Hydrogen peroxide.
☐ A baby food jar.
☐ A Popsicle stick.
☐ Matches (a pilot candle).
☐ A disposal can for matches.

**(TO)** study the slow oxidation of iron into rust. To monitor the consumption of oxygen in this process.

---

## IRON RUSTS         O         Oxidation (    )

1. Fluff a dime-sized pinch of steel wool into a light "cloud." Push it to the bottom of a test tube with your pencil.

OXIDIZER
STEEL WOOL

2. Add 3 droppersful of oxidizer to the test tube and stand it in a small jar. Watch iron (chemical symbol Fe) oxidize to form rust:

$$2\,Fe + O_2 \longrightarrow 2\,FeO + energy$$
iron    oxygen gas      iron oxide (solid rust)

    a. What evidence do you see *and* feel that oxidation is happening?
    b. Besides FeO, iron also oxidizes to $Fe_2O_3$ and $Fe_3O_4$. Write balanced equations for these other 2 forms of rust.

3. Rubberband a dry test tube upside-down to the side of a battery. Force a half clothespin between so the tube stands inverted in a lid full of water.

EMPTY (CONTROL)        STEEL WOOL

4. Fill the steel-wool test tube *to the top* with water, then pour it down a sink. Shake out all liquid, then invert this test tube like the first in the same lid.

5. Set the lid and test tubes aside for 24 hours. Predict what changes (if any) you'll see in each test tube after 24 hours.

© 1995 by TOPS Learning Systems       10

---

## Introduction

Matter takes up much more space as a gas than as a liquid or solid. Illustrate this concept with a graduated cylinder filled with 10 mL of water. Discuss what volume this much water occupies in each phase:
- solid: slightly more than 10 mL. (Ice floats on water.)
- liquid: 10 mL. This volume is now in the cylinder.
- gas: 22.4 L, almost 6 gallons! (This is the volume occupied by 1 gram molecular weight of any gas at standard temperature and pressure.)

## Answers / Notes

2a. The colorless oxidizer rapidly turns rust red. The bottom of the test tube becomes warm (compared to the cooler top), indicating that the reaction produces heat energy.

2b.
$$4\,Fe + 3\,O_2 \longrightarrow 2\,Fe_2O_3 + energy$$
$$3\,Fe + 2\,O_2 \longrightarrow Fe_3O_4 + energy$$

*While these equations are correct in principle, they leave out water, an essential yet complicating ingredient in the chemistry of rust. Here is a closer approximation of reality:*

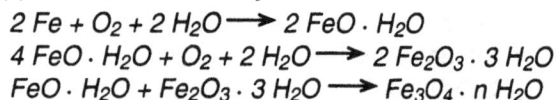

$$2\,Fe + O_2 + 2\,H_2O \longrightarrow 2\,FeO \cdot H_2O$$
$$4\,FeO \cdot H_2O + O_2 + 2\,H_2O \longrightarrow 2\,Fe_2O_3 \cdot 3\,H_2O$$
$$FeO \cdot H_2O + Fe_2O_3 \cdot 3\,H_2O \longrightarrow Fe_3O_4 \cdot n\,H_2O$$

4. *Filling the test tube to the top, and then pouring the water out, completely exchanges oxygen-depleted air inside the test tube with fresh room air. It also safely disposes of the skin-irritating oxidizer. The steel wool should remain at the bottom of the test tube.*

5. After 24 hours, steel wool in the experiment tube should turn rusty red. Water will rise inside the tube, replacing oxygen gas that combined with iron to form solid rust. No changes should happen in the control tube, because it contains no steel wool.

## Materials

☐ A graduated cylinder to use in the introduction (optional).
☐ Pure steel wool balls. Avoid presoaped pads. If you have a choice, select the finest (thinnest) grade possible.
☐ Two small or medium-sized test tubes of equal size. (Reserve your largest size, if available, for making oxygen in the next activity.)
☐ A dropper bottle labeled "Oxidizer." Combine 1 part vinegar to 2 parts chlorine bleach. Do not substitute a non-chlorine bleach like hydrogen peroxide. It won't oxidize the steel wool rapidly enough to allow immediate identification of rust.
☐ A baby food jar.
☐ Two rubber bands.
☐ A clothespin (two clothespin halves).
☐ Two size-D batteries, dead or alive.
☐ A large jar lid, canning size or larger.

**(TO)** understand that air is a mixture of gases. To estimate the percent, by volume, of oxygen in the air we breathe.

## PERCENT OXYGEN?　　　　O　　　　Oxidation (　　)

1. Mark the water level inside each test tube from the last activity (as necessary) with slivers of masking tape. (Do this *before* moving the experiment back to your work area.)

    a. Evaluate your hypothesis in step 5 of the last activity.
    b. Save your rusty steel wool in a small jar to dry out. You'll need it in activity 13.

TAPE POINTER

2. Assuming that all oxygen inside the test tube is now locked up in rust, what percentage of the original air was oxygen? Show your math.

3. About 99% of the air we breath is composed of oxygen and nitrogen.

$O_2$ + $N_2$ + other gases
99%    1%

    a. What percent is made from nitrogen?
    b. What gases might be in the remaining 1%?

4. Think of a way to make water rise much higher in a test tube of rusting steel wool. Detail what you did and what you learned.

11

## Answers / Notes

1. *The level of water in the steel-wool test tube needs to be marked before moving, or air could get inside. The control needs no marker, since no water has entered.*

1a. *Students should confirm that their previous predictions were correct, or explain, in terms of new evidence, why they were wrong.*

2. *These calculations are valid only for test tubes of uniform diameter, where length is proportional to volume. Though measurements will vary according to the length of test tubes used, the ratio should remain close to 0.21. (Higher ratios may be caused by excess water initially trapped in the steel wool; lower ratios by oxygen-depleted air trapped in the tube before it was inverted in water.)*

$$\% \ O_2 = \frac{\text{height water}}{\text{height tube}} = \frac{26 \text{ mm}}{122 \text{ mm}} = .021 \times 100 = 21\%$$

3a. nitrogen = 99% − 21% = 78%
3b. Room air contains carbon dioxide and water vapor. These gases must be part of this last 1% *(as well as trace amounts of neon and other noble gases).*

4. Procedure: (A) Prepare a test tube of steel wool as in task 10, and invert it in a large tube of water. *(Rinsing it first with oxidizer is an optional procedure that will speed the rusting process.)*
(B) Prepare oxygen in a larger test tube as in task 9.

Transfer it to the smaller, steel-wool test tube by underwater displacement as in task cards 3 and 7. (C) Stand the inverted tube of steel wool and oxygen in a small jar of water. Monitor the rising water inside with a pointer of masking tape. *(The lid and battery arrangement of task 10 is not needed, because water rises high enough inside the tube for easy observation.)*

Results: *(It may take up to 48 hours for all oxygen in the test tube to be consumed, and the water level to stop rising. The height it reaches depends on the purity of the original oxygen trapped inside. Here is one result.)*

$$\% \ O_2 = \frac{\text{height water}}{\text{height tube}} = \frac{99 \text{ mm}}{122 \text{ mm}} = .081 \times 100 = 81\%$$

## Materials

☐ The previous experiment in progress, with write-up.
☐ Masking tape and scissors.
☐ A baby food jar to hold wet, rusty steel wool.
☐ A metric ruler.
☐ A calculator (optional).
☐ Steel wool and a dropper bottle of oxidizer.
☐ Yeast and hydrogen peroxide.
☐ A large test tube for making oxygen.
☐ A smaller test tube to hold the steel wool, 3/4 the volume of the larger tube or less. Using test tubes of equal size, you will need 2 tubes of oxygen to fully displace all the water in 1 test tube of steel wool.
☐ A tub of water, wider than your longest test tube.

**(TO)** study the rapid oxidation of iron to iron oxide. To identify variables that speed this burning process.

---

## DOES IRON BURN?     O        Oxidation (   )

1. Clamp each object below in a clothespin. Try to ignite it with a burning candle that stands on a sheet of paper. Does each object glow, or burn, or what?

   a. A nail: ◄━━━━━━━

   b. Steel wool tightly rolled into a "snake:"

   c. A thin "cloud" of steel wool teased apart:

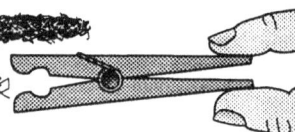

2. Repeat step 1b, this time lowering the glowing "snake" into a test tube of oxygen. What happens?

3. Identify 2 ways to make iron burn.

4. When iron burns, it unites with oxygen to form magnetic iron oxide ($Fe_3O_4$).

   a. Write a balanced equation.
   b. Describe the color and shape of this oxide. Does it look anything like rusty $Fe_3O_4$?
   c. Use a magnet to verify that this oxide really is magnetic.

© 1995 by TOPS Learning Systems       12

---

## Answers / Notes

1a. The nail collects black soot (from the incomplete combustion of paraffin to carbon dioxide and water vapor). But it neither glows nor burns.

1b. The tightly rolled steel-wool "snake" glows red hot, but very little is consumed by combustion.

1c. The "cloud" of steel wool ignites in the candle flame and continues to burn when you remove it from the heat! The thicker strands of iron glow red hot, while the thinner strands burn down like tiny fuses. Tiny spherical lumps of matter remain after the iron burns. Some of these spheres remain attached to their original strands, while others drop onto the paper and burn tiny holes in it.

   *This paper surface serves a dual purpose: It protects your table top, and gives a dramatic visual warning not to touch the tiny balls of iron oxide before they cool.*

2. The "snake" glows dull red in ordinary air, but burns white hot when plunged into a test tube of oxygen gas.

3. To make iron burn…
   (a) Divide it into very fine pieces surrounded by ample air. *(Maximize the ratio of surface area to total volume. Iron ground to a fine powder and blown over a candle flame will burn intensely.)*
   (b) Surround it with higher concentrations of oxygen.

4a. $3\,Fe + 2\,O_2 \longrightarrow Fe_3O_4 + energy$

4b. This iron oxide is composed of many tiny lumps that are grey-black, spherical and hard. It looks nothing like red, flaky, soft rust. *(These oxides look different because the rusty form is hydrated, while the burned form is not. Students will discover this in the next activity. For now, let the problem remain unresolved.)*

4c. These tiny balls of $Fe_3O_4$ are attracted by a magnet just like the original steel wool.

## Materials

- ☐ A candle, birthday-size or larger, standing in a lump of clay.
- ☐ A sheet of scratch paper.
- ☐ Matches (a pilot candle).
- ☐ A disposal can for matches.
- ☐ A clothespin.

- ☐ A nail.
- ☐ Steel wool.
- ☐ A test tube.
- ☐ Hydrogen peroxide.
- ☐ Yeast.
- ☐ A magnet.

**(TO)** distinguish between hydrated and anhydrous iron oxide. To derive a balanced chemical equation to show how iron oxides combine when heated.

---

## CHEMISTRY PUZZLE     O     Oxidation (    )

1. Get a small strip of aluminum foil. Press one end gently into a washer to form a small "spoon."

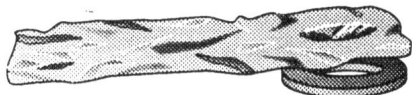

2. Get the rusty steel wool that you previously set aside. Put a tiny pinch of rust in your foil spoon, and heat it over a candle flame. What happens?

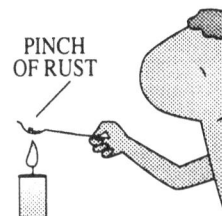

PINCH OF RUST

3. Red iron oxides are hydrated (contain water); black iron oxides are anhydrous (contain no water). Why did these oxides change color when you heated them?

4. Collect more rust on a piece of white paper by gently rubbing the rusty steel wool between your fingers.
    a. List the 3 hydrated forms of iron oxide now on the paper.
    b. Remove all $Fe_3O_4$ so only 2 oxides remain. Explain how you did this.

5. Make a new foil spoon.
    a. Heat the remaining oxide mixture over a candle flame, then test with a magnet. What happens?
    b. Propose a balanced chemical reaction that shows how these iron oxides combine.

© 1995 by TOPS Learning Systems     13

---

## Answers / Notes

2. The rust-red oxides turn grey-black when heated by the candle flame.

3. These iron oxides changed color because candle heat drove off the water of hydration.

4a. *These 3 hydrated oxides were identified in activity 10:*   $FeO$, $Fe_2O_3$ and $Fe_3O_4$.

4b. In activity 12, $Fe_3O_4$ was shown to be magnetic. To remove this particular form of iron oxide from the mixture, simply pass a magnet back and forth over the paper, until all of the $Fe_3O_4$ sticks to it.

5a. Both $FeO$ and $Fe_2O_3$ changed from red to black as candle heat drove of their water of hydration. In addition, these previously nonmagnetic oxides are now attracted by a magnet.

5b. Both red hydrated iron oxides combined to form a grey-black anhydrous magnetic oxide:

$$FeO + Fe_2O_3 \longrightarrow Fe_3O_4$$

*(Residual unmagnetized oxide is anhydrous $Fe_2O_3$, which was initially excess hydrated $Fe_2O_3$, or excess hydrated FeO that was oxidized to $Fe_2O_3$:*   $4\ FeO + O_2 \longrightarrow 2\ Fe_2O_3$.)

## Materials

☐ Aluminum foil.
☐ A washer with a hole as large as a fingertip.
☐ Rusty steel wool that students set aside in activity 11.
☐ A birthday-size candle or larger, with clay support base.
☐ Matches (a pilot candle).
☐ A disposal can for matches.
☐ A magnet.

notes 13        enrichment

**(TO)** hypothesize why a burning candle creates a partial vacuum in an inverted jar. To appreciate that this, and all hypotheses, are subject to change based on new experimental evidence.

---

## WHAT'S YOUR HYPOTHESIS? ◯       Oxidation (   )

1. Use clay to stand a birthday candle in the center of a plastic lid. Light the candle and cover it with a large jar.

   a. After the flame dies, pick up the jar. What happens?
   b. Propose an *hypothesis* to explain your observations.
   c. Purge the jar of smoky air by filling it with water, then emptying it.

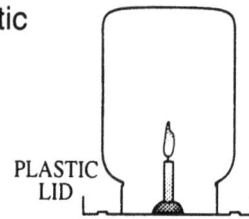

PLASTIC LID

2. Stand your birthday candle in a shallow bowl of water. Light it and cover with your large jar full of fresh air.

   a. What happens after the candle goes out?
   b. Do these results *support* your hypothesis in 1b?
   c. Purge the jar of smoky air. If the wick is wet, squeeze it dry in a paper towel.

3. Mix a drop of liquid soap into the bowl of water. Stand a lighted candle in it and cover with your pint jar as before.

   a. What do the bubbles mean?
   b. Revise your hypothesis in 1b on the basis of this new evidence.

4. A scientific hypothesis is always tentative. What does this mean?

14

---

### Answers / Notes

1a. The lid sticks to the jar when you pick it up. Pry off the lid, and air rushes in to fill the partial vacuum. *(If no vacuum forms, check that the lid and jar rim make smooth contact.)*

1b. Allow all reasoned hypotheses to stand. Here is one example: the burning candle creates a partial vacuum because it removes oxygen from the jar, while no new air can enter to take its place. *(Step 3 will reveal new experimental evidence that calls this particular hypothesis into question.)*

2a. Water rises in the inverted jar. *(There are other things to see as well: (a) Air bubbles out of the jar before water rises inside. (b) Water rises in the jar only after the candle goes out. These more subtle observations will likely be lost on all but your most alert students. That's OK.)*

2b. Allow all reasoned answers to stand. *(Students who believe that the vacuum was created by the consumption of oxygen, who only observed that water rose inside the jar and nothing more, will logically conclude that their hypothesis is confirmed. They may still argue that water rose inside the jar to replace oxygen that was consumed by the candle flame.)*

3a. The bubbles show that air exits the jar before water enters.

3b. Oxygen consumption does not entirely account for water rising inside the jar. There are other considerations: The candle flame heats and expands air inside the jar so that some of it is pushed out. After the flame dies, air that remains inside cools and contracts, pulling in water as it takes up less space.

4. A scientific hypothesis is our best explanation based on current experimental evidence. It is tentative in nature because new observations may contradict what we thought was true and require us to modify our views.

---

### Materials

- ☐ A birthday candle with a clay base.
- ☐ The plastic lid from a margarine tub or equivalent container.
- ☐ A quart jar.
- ☐ Matches (a pilot candle).

- ☐ A disposal can for matches.
- ☐ A shallow bowl.
- ☐ Liquid soap.
- ☐ A paper towel.

**(TO)** repeat an experiment with uncontrolled variables. To appreciate that its experimental results are not reproducible.

---

## REPRODUCIBLE RESULTS?   O                    Oxidation (    )

1. Run masking tape ⅓ the length of a graduated cylinder, starting near the rim.

2. Stand a birthday candle in clay and set it in a shallow bowl of soapy water. Light it and turn your taped cylinder upside down over it.

    a. Mark the tape to record how high the water rises.
    b. Purge the cylinder of smoky air. Blot the wet candle wick in a paper towel to dry it.

3. Repeat step 2 the same way at least 3 more times.
    a. Are your results *reproducible*? (Does water rise to the same height each time?)
    b. What part of step 2 is difficult to *control* (repeat exactly the same way each time)?

4. In activities 10-11, you put steel wool in an inverted test tube, and calculated the amount of oxygen in room air. Why is such a calculation valid in the steel wool experiment, but not in this candle experiment?

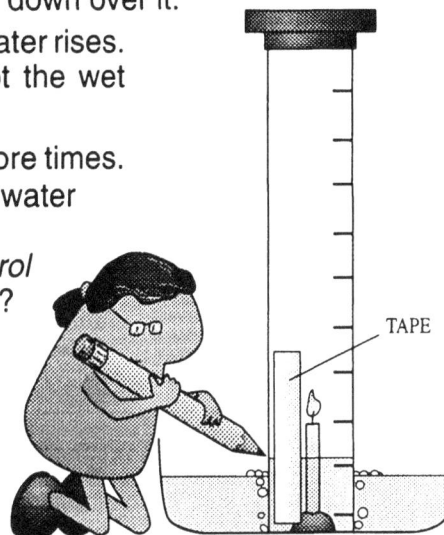

TAPE

15

## Answers / Notes

3a. No. Water rises to different levels when the experiment is repeated. A wide distribution of pencil marks indicates that these experimental results are not reproducible.

3b. It is difficult to control how quickly you submerge the mouth of the cylinder under water when you place it over the candle. *(Doing this as fast as possible seems to maximize how high the water rises.)*

4. *This is an open-ended question that invites a response at many levels. Encourage thoughtful consideration in any or all of these areas:*

• The steel-wool experiment had reproducible results based on one controlled variable. *(Though it was not repeated by individual lab groups, different lab groups got similar results.)* The candle experiment, by contrast, turned out differently each time, with important variables remaining uncontrolled.

• Steel wool rusts (oxidizes) slowly. Heat was not produced so fast that air expanded and escaped the test tube. Oxygen was not consumed so rapidly that the speed with which you inverted the test tube over water was a significant variable.

• Candles burn (oxidize) rapidly. High variable temperatures inside the cylinder expanded the air by different volumes. Some of this hot air escaped before it could be trapped in the soap bubbles.

• The steel wool continued to oxidize until all oxygen was consumed and the water level rose no higher. The candle only burned oxygen near its flame, but likely failed to consume all the oxygen in the cylinder before the flame died.

• Steel wool and oxygen produce solid oxides of iron that take up very little space. Candle wax and oxygen produce products of greater volume:

$$2\,C_{28}H_{58} + 85\,O_2 \longrightarrow 56\,CO_2 + 58\,H_2O + Energy$$

For every 85 molecules of oxygen consumed, 56 molecules of carbon dioxide and 58 molecules of water vapor are produced. The water vapor likely condenses to a small volume, but carbon dioxide remains, replacing $56/85$ of the volume lost to the consumption of oxygen.

## Materials

☐ A 100 mL graduated cylinder.
☐ Masking tape.
☐ A shallow bowl of soapy water.
☐ A birthday candle set in a small clay base.
☐ Matches (a pilot candle).
☐ A disposal can for matches.
☐ A paper towel.

**(TO)** write a balanced equation, based on experimental evidence, for the combustion of alcohol.

## ISOPROPYL INVESTIGATION ○        Oxidation ( )

1. Deposit a dropperful of 70% isopropyl alcohol in the bottom "dish" of a small inverted jar. Light it with a match and write your observations.

> **CAUTION:** glass held in the flame gets <u>much</u> hotter than the glass holding the burning alcohol, even though they both look cool. To avoid blistering burns, exercise caution and good judgement at all times.

2. Carefully perform tests on this chemical reaction to answer the following questions:
   - a. Is this an oxidation reaction?
   - b. Is water vapor produced?
   - c. Is carbon dioxide produced?

3. Look up the formula for isopropyl alcohol in the dictionary. Write a balanced equation for the combustion of this substance.

16

## Answers / Notes

*Have you ever wished your science tests could go beyond paper and pencil? Consider using this final activity as a testing instrument. It will allow students to demonstrate their mastery of important ideas within the context of real laboratory experimentation.*

*Seventy percent rubbing alcohol is relatively safe to use, because its 30% water content keeps combustion temperatures relatively low. Spill the burning liquid onto a piece of paper and it will not ignite. (You can even spill the fiery liquid on your palm without blistering the skin). This does not imply that the flame itself is cool. It burns blue hot, hotter than a candle flame. Hold paper above this flame and it will quickly ignite.*

*A greater, more subtle danger is that students who work with this relatively safe liquid will over-generalize and develop a cavalier attitude about burning extremely dangerous hydrocarbons like gasoline. Reduce this danger by tolerating absolutely no horseplay or unauthorized experimentation with burning rubbing alcohol.*

1. The alcohol burns vigorously with a blue flame at its surface and a yellow flame above. A residue of liquid remains after the flame goes out. *(A 70% alcohol solution contains 30% water, which remains unoxidized.)*

2a. Yes, this is an oxidation reaction. Cover the flame with another small jar, and the flame dies. You have cut off its supply of oxygen.

2b. Yes, water vapor is produced. Cover the flame with another cool jar, and moisture condenses inside.

2c. Yes, carbon dioxide is produced. Trap combustion gases in another small jar by covering the flame. *(Caution. Avoid holding the mouth of the jar in the flame long enough to overheat. Just place it quickly over the flame.)* Add limewater to these gases, and shake to mix. A milky white precipitate will form, indicating the presence of carbon dioxide.

3.    $2\,C_3H_8O + 9\,O_2 \longrightarrow 6\,CO_2 + 8\,H_2O + energy$

## Materials

☐ 70% isopropyl alcohol (rubbing alcohol), sold in drug stores. 70% ethyl alcohol may be substituted. NEVER substitute 100% pure alcohol, gasoline or other highly flammable hydrocarbons.

☐ Two baby food jars.

☐ A dropper bottle for the alcohol. Or dispense it in a third baby food jar with an eye dropper.

☐ Matches (a pilot candle).

☐ A disposal can for matches.

☐ Limewater.

☐ A dictionary.

# REPRODUCIBLE MATERIALS

## Task Cards

# Task Cards Options

**Here are 3 management options to consider before you photocopy:**

**1. Consumable Worksheets:** Copy 1 complete set of task card pages. Cut out each card and fix it to a separate sheet of boldly lined paper. Duplicate a class set of each worksheet master you have made, 1 per student. Direct students to follow the task card instructions at the top of each page, then respond to questions in the lined space underneath.

**2. Nonconsumable Reference Booklets:** Copy and collate the 2-up task card pages in sequence. Make perhaps half as many sets as the students who will use them. Staple each set in the upper left corner, both front and back to prevent the outside pages from working loose. Tell students that these task card booklets are for reference only. They should use them as they would any textbook, responding to questions on their own papers, returning them unmarked and in good shape at the end of the module.

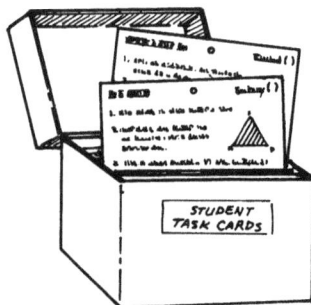

**3. Nonconsumable Task Cards:** Copy several sets of task card pages. Laminate them, if you wish, for extra durability, then cut out each card to display in your room. You might pin cards to bulletin boards; or punch out the holes and hang them from wall hooks (you can fashion hooks from paper clips and tape these to the wall); or fix cards to cereal boxes with paper fasteners, 4 to a box; or keep cards on designated reference tables. The important thing is to provide enough task card reference points about your classroom to avoid a jam of too many students at any one location. Two or 3 task card sets should accommodate everyone, since different students will use different cards at different times.

## CANDLE COMBUSTION    ◯    Oxidation (    )

1. Stand 2 birthday candles in small lumps of clay. Put one next to a small jar, the other next to a large jar.

READY?

2. Light the candles, then set a jar over each one at the same time. Write your observations.

3. Blow fresh air into each jar, and repeat the experiment. Time how long each candle burns inside each jar.

4. *Combustion* (burning) is a process that uses up both fuel and oxygen.
   a. Compare a new candle with a used one. What fuel is being consumed?
   b. Why do burning times vary with jar size?

5. When a fuel burns, it *oxidizes* (combines with oxygen).
   a. What gets oxidized in this experiment?
   b. Does this oxidation reaction absorb energy or release energy? Explain.

© 1995 by TOPS Learning Systems                    1

---

## HUMAN RESPIRATION    ◯    Oxidation (    )

SANDWICH BAG

1. Cut an empty toilet paper roll into 2 equal tubes. Rubberband the mouth of a sandwich bag around one, a plastic produce bag around the other.

PRODUCE BAG

2. Take a deep breath and hold your nose. Breathe in and out through your mouth, as normally as possible, into the larger bag.
   a. Look at a clock. How long can you use the same air over and over before you feel uncomfortable?
   b. How did your breathing change over time?
   c. Did you collect anything in the bag besides "used" air?

3. Now exhale fully into the small bag, allowing excess air to leak past the mouthpiece. Once again, hold your nose and breathe through your mouth as normally as possible.
   a. How long did you rebreathe the same air? Compare this result with step 2a.
   b. Compare human *respiration* (breathing) to candle combustion.

4. A candle oxidizes wax.
   a. What do you think *you* oxidize?
   b. Does the oxidation reaction in your body absorb or release energy? Explain.

© 1995 by TOPS Learning Systems                    2

## I FEEL FAINT     ⚪     Oxidation (    )

1. Wrap masking tape around the end of a rubber tube to make a sanitary mouthpiece. You will blow air through it.

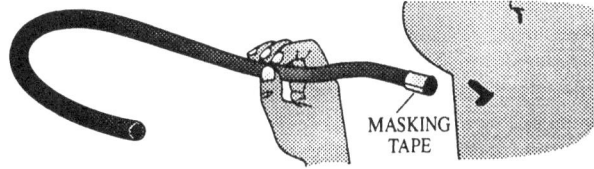

MASKING TAPE

2. Devise a way to collect and seal only your breath (and no other air) inside a pint jar. Hint: use these materials.

   a. Detail your method in words and pictures.

   b. Hold your breath as long as you can, then collect it inside a pint jar *before* you draw a breath of fresh air. Close with a lid.

BIG BREATH (Hold as long as possible)

TUB OF WATER

TUBE

PINT JAR WITH LID

3. Show that this pint of "breathed" air contains little or no oxygen. Explain your method and results.

3

---

## BOTTOM BURNER     ⚪     Oxidation (    )

1. Loop the end of a 30 cm piece of wire around a pencil. Press the clay base of a birthday candle into this loop.

2. Shape this wire so the candle stands centered at the bottom of a small jar, while the other end of the wire is looped and rubber-banded to a battery "handle."

3. Use the handle to lift the candle from the jar, light it, and lower it back inside the uncovered jar. Does it keep burning?

4. Hold another small jar, mouth to mouth, over the first.
   a. What happens inside the top jar?
   b. What happens inside the bottom jar?

5. Draw a side view of the candle and jar, showing how *convection currents* (air streams driven by heat) supply a steady flow of oxygen to the candle flame. Label and explain your drawing.

WIRE LOOP

6. Blow your breath into a cool, dry jar and observe closely. Name 2 ways you are like a burning candle. (Save your candle holder to use again.)

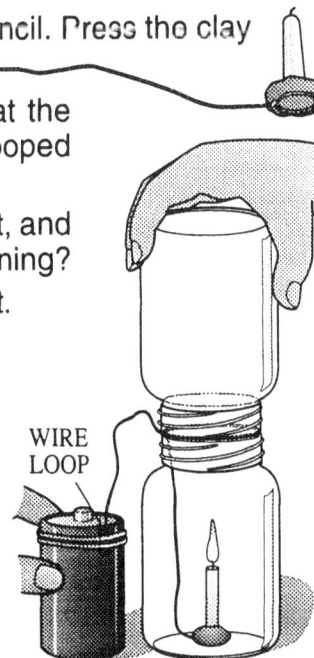

4

## IT'S A GAS  ◯  Oxidation (   )

1. Add 3 tablespoons (36 mL) vinegar to a pint jar. Dry the spoon, then add 1 rounded tablespoon (18 mL) of baking soda. Cover the mouth of the jar with an index card.

2. What evidence do you see that this chemical reaction is taking place: **Vinegar + Baking Soda ⟶ Carbon Dioxide Gas**

3. Use your candle holder and battery to hold a short burning candle in the bottom of a small jar, as before.

4. Pour carbon dioxide gas (not the liquid underneath) over your burning candle. Explain your observations in terms of one gas *displacing* (taking the place of) another.

5. Add a pinch of baking powder to half a test tube of vinegar. Cap it tightly with your thumb and invert over a sink.
   a. What happens? Why?
   b. Design a vinegar and baking soda fire extinguisher. Include operating instructions with your design.

5

---

## LIMEWATER REACTION  ◯  Oxidation (   )

1. Fill 2 small jars about 1/4 full with limewater. Label one C for Control, the other E for Experiment.

2. Rubberband the mouth of your Control with a small square of plastic wrap. Reserve another rubber band and square of wrap for step 3.

3. Make a pint of carbon dioxide as before. Pour all this gas (not the liquid underneath) into your Experiment, then seal it as quickly as possible.

4. Describe what happens (if anything) in each jar.

5. Interpret your observations in terms of this equation:

**Limewater + Carbon Dioxide Gas ⟶ Calcium Carbonate**

PRECIPITATE (a white solid)

6. Your experiment and control are alike in every respect but one.
   a. What is the *controlled variable* (the one difference)?
   b. Why have a control?

7. Remove the plastic wrap from both jars and set them aside overnight. How does the limewater surface change over time? What can you conclude?

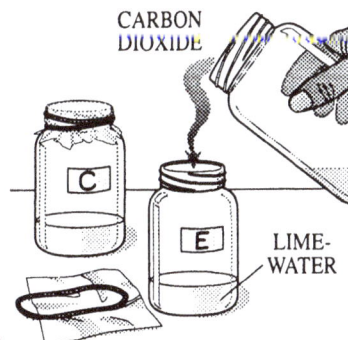

6

## THREE LITTLE JARS  ○  Oxidation (   )

1. Rinse 3 small jars of equal size with water. Cover each jar with a labeled lid identifying the air you put inside.

**a. ROOM AIR:**

It's already in the jar.

**b. RESPIRATION AIR:**

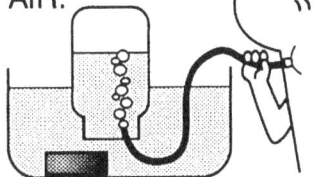

Hold your breath as long as you can. Collect it by water displacement.

**c. COMBUSTION AIR:**

Light a candle in its holder and set it in the bottom of another jar. Hold your test jar inverted over the top until the flame goes out.

2. These jars are all the same except for one controlled variable. What is it?

3. Test the gas in each jar for carbon dioxide: open each lid just wide enough to pour in limewater, close it quickly, and shake the jar.

    a. Report your results.    b. What can you conclude?

7

---

## CHEMICAL OVERVIEW  ○  Oxidation (   )

1. Review the (**activity**/step) under each blank, then write the correct word.

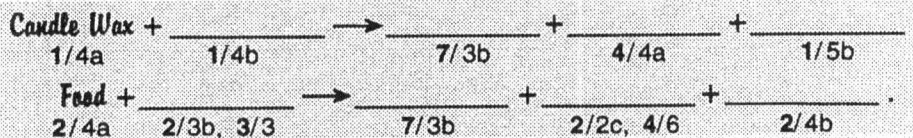

Candle Wax + _____ ⟶ _____ + _____ + _____ .
   1/4a     1/4b       7/3b     4/4a     1/5b

Food + _____ ⟶ _____ + _____ + _____ .
  2/4a  2/3b, 3/3    7/3b   2/2c, 4/6   2/4b

2. Pair each substance with its correct molecular formula:

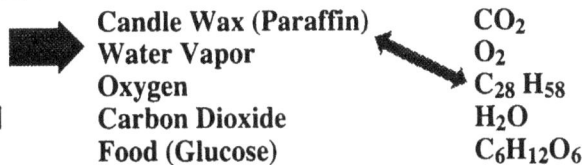

3. Rewrite both equations in step 1 using molecular formulas.

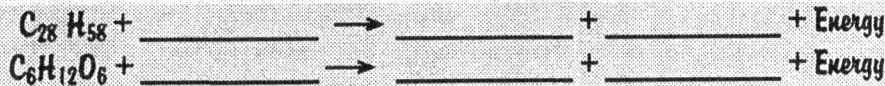

Candle Wax (Paraffin)    $CO_2$
Water Vapor    $O_2$
Oxygen    $C_{28}H_{58}$
Carbon Dioxide    $H_2O$
Food (Glucose)    $C_6H_{12}O_6$

$C_{28}H_{58}$ + _____ ⟶ _____ + _____ + Energy
$C_6H_{12}O_6$ + _____ ⟶ _____ + _____ + Energy

4. Rewrite both equations in step 3 so they balance. Each side must have the same number of <u>C</u>arbon atoms, <u>H</u>ydrogen atoms, and <u>O</u>xygen atoms.

YOU **CAN** CHANGE THE NUMBER OF MOLECULES...

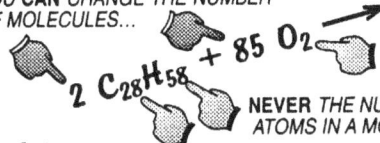

$2\ C_{28}H_{58}$ + $85\ O_2$

...BUT **NEVER** THE NUMBER OF ATOMS IN A MOLECULE.

8

## GLOWING SPLINT  O  Oxidation (   )

1. A molecule of hydrogen peroxide is like water, but with 1 extra oxygen atom. Write the formula for both substances.

2. Hydrogen peroxide decomposes (breaks down) to form water and oxygen gas. Write a *balanced* equation (same number of atoms on both sides of arrow).

3. Put a *small* pinch of yeast in the bottom of a clean, dry test tube.
   a. Fill the test tube ¼ full with <u>hydrogen peroxide</u>. Rest it in a small jar.
   b. Describe and interpret the reaction (if any).
   c. Burn a Popsicle stick for about 10 seconds. Blow out the flame, then poke the glowing splint halfway into the test tube. What happens?

4. Set up a control by repeating all of step 3, but substitute <u>water</u> for <u>hydrogen peroxide</u> in step 3a. This is your controlled variable.
   a. What happens differently?
   b. Does the glowing splint reveal that the control test tube contained at least *some* oxygen? Explain.

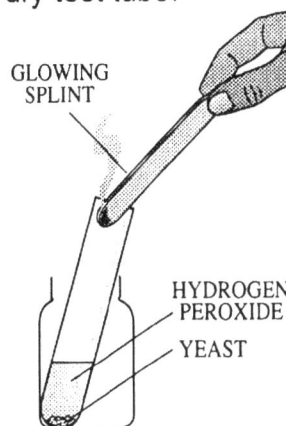

GLOWING SPLINT

HYDROGEN PEROXIDE

YEAST

© 1995 by TOPS Learning Systems                                        9

---

## IRON RUSTS  O  Oxidation (   )

1. Fluff a dime-sized pinch of steel wool into a light "cloud." Push it to the bottom of a test tube with your pencil.

OXIDIZER

STEEL WOOL

2. Add 3 droppersful of oxidizer to the test tube and stand it in a small jar. Watch iron (chemical symbol Fe) oxidize to form rust:

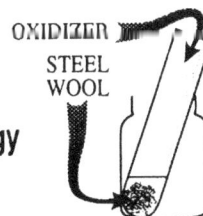

$$2\,Fe + O_2 \longrightarrow 2\,FeO + energy$$

iron   oxygen   iron oxide
       gas      (solid rust)

   a. What evidence do you see *and* feel that oxidation is happening?
   b. Besides FeO, iron also oxidizes to $Fe_2O_3$ and $Fe_3O_4$. Write balanced equations for these other 2 forms of rust.

3. Rubberband a dry test tube upside-down to the side of a battery. Force a half clothespin between so the tube stands inverted in a lid full of water.

EMPTY (CONTROL)

STEEL WOOL

4. Fill the steel-wool test tube *to the top* with water, then pour it down a sink. Shake out all liquid, then invert this test tube like the first in the same lid.

5. Set the lid and test tubes aside for 24 hours. Predict what changes (if any) you'll see in each test tube after 24 hours.

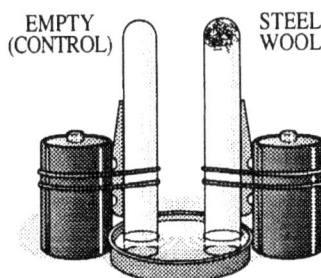

© 1995 by TOPS Learning Systems                                        10

## PERCENT OXYGEN?    ○    Oxidation (    )

1. Mark the water level inside each test tube from the last activity (as necessary) with slivers of masking tape. (Do this *before* moving the experiment back to your work area.)

    a. Evaluate your hypothesis in step 5 of the last activity.

    b. Save your rusty steel wool in a small jar to dry out. You'll need it in activity 13.

TAPE POINTER

2. Assuming that all oxygen inside the test tube is now locked up in rust, what percentage of the original air was oxygen? Show your math.

3. About 99% of the air we breath is composed of oxygen and nitrogen.

$O_2 + N_2$ + other gases
99%       1%

    a. What percent is made from nitrogen?

    b. What gases might be in the remaining 1%?

4. Think of a way to make water rise much higher in a test tube of rusting steel wool. Detail what you did and what you learned.

11

## DOES IRON BURN?    ○    Oxidation (    )

1. Clamp each object below in a clothespin. Try to ignite it with a burning candle that stands on a sheet of paper. Does each object glow, or burn, or what?

    a. A nail:

    b. Steel wool tightly rolled into a "snake:"

    c. A thin "cloud" of steel wool teased apart:

2. Repeat step 1b, this time lowering the glowing "snake" into a test tube of oxygen. What happens?

3. Identify 2 ways to make iron burn.

4. When iron burns, it unites with oxygen to form magnetic iron oxide ($Fe_3O_4$).

    a. Write a balanced equation.

    b. Describe the color and shape of this oxide. Does it look anything like rusty $Fe_3O_4$?

    c. Use a magnet to verify that this oxide really is magnetic.

12

## CHEMISTRY PUZZLE ○     Oxidation (  )

1. Get a small strip of aluminum foil. Press one end gently into a washer to form a small "spoon."

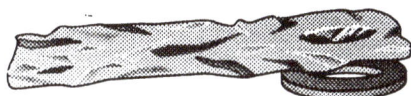

2. Get the rusty steel wool that you previously set aside. Put a tiny pinch of rust in your foil spoon, and heat it over a candle flame. What happens?

PINCH OF RUST

3. Red iron oxides are hydrated (contain water); black iron oxides are anhydrous (contain no water). Why did these oxides change color when you heated them?

4. Collect more rust on a piece of white paper by gently rubbing the rusty steel wool between your fingers.
    a. List the 3 hydrated forms of iron oxide now on the paper.
    b. Remove all $Fe_3O_4$ so only 2 oxides remain.
Explain how you did this.

5. Make a new foil spoon.
    a. Heat the remaining oxide mixture over a candle flame, then test with a magnet. What happens?
    b. Propose a balanced chemical reaction that shows how these iron oxides combine.

13

---

## WHAT'S YOUR HYPOTHESIS? ○     Oxidation (  )

1. Use clay to stand a birthday candle In the center of a plastic lid. Light the candle and cover it with a large jar.
    a. After the flame dies, pick up the jar. What happens?
    b. Propose an *hypothesis* to explain your observations.
    c. Purge the jar of smoky air by filling it with water, then emptying it.

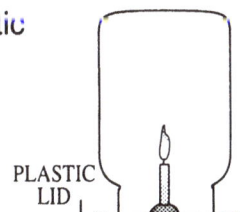

PLASTIC LID

2. Stand your birthday candle in a shallow bowl of water. Light it and cover with your large jar full of fresh air.
    a. What happens after the candle goes out?
    b. Do these results *support* your hypothesis in 1b?
    c. Purge the jar of smoky air. If the wick is wet, squeeze it dry in a paper towel.

3. Mix a drop of liquid soap into the bowl of water. Stand a lighted candle in it and cover with your pint jar as before.
    a. What do the bubbles mean?
    b. Revise your hypothesis in 1b on the basis of this new evidence.

4. A scientific hypothesis is always tentative. What does this mean?

14

## REPRODUCIBLE RESULTS?   O                    Oxidation (    )

1. Run masking tape ⅓ the length of a graduated cylinder, starting near the rim.

2. Stand a birthday candle in clay and set it in a shallow bowl of soapy water. Light it and turn your taped cylinder upside down over it.

   a. Mark the tape to record how high the water rises.
   b. Purge the cylinder of smoky air. Blot the wet candle wick in a paper towel to dry it.

3. Repeat step 2 the same way at least 3 more times.

   a. Are your results *reproducible*? (Does water rise to the same height each time?)
   b. What part of step 2 is difficult to *control* (repeat exactly the same way each time)?

4. In activities 10-11, you put steel wool in an inverted test tube, and calculated the amount of oxygen in room air. Why is such a calculation valid in the steel wool experiment, but not in this candle experiment?

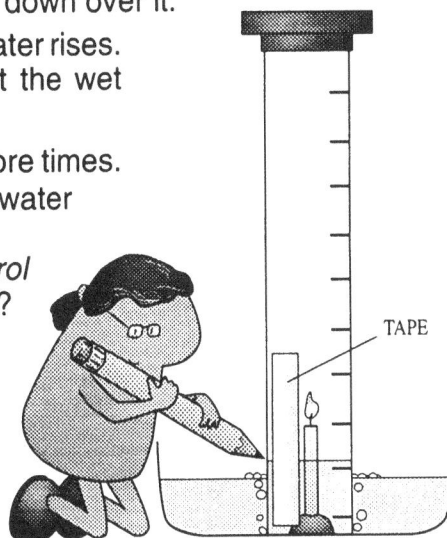

TAPE

15

## ISOPROPYL INVESTIGATION   O                    Oxidation (    )

1. Deposit a dropperful of 70% isopropyl alcohol in the bottom "dish" of a small inverted jar. Light it with a match and write your observations.

**CAUTION:** glass held in the flame gets <u>much</u> hotter than the glass holding the burning alcohol, even though they both look cool. To avoid blistering burns, exercise caution and good judgement at all times.

2. Carefully perform tests on this chemical reaction to answer the following questions:

   a. Is this an oxidation reaction?
   b. Is water vapor produced?
   c. Is carbon dioxide produced?

3. Look up the formula for isopropyl alcohol in the dictionary. Write a balanced equation for the combustion of this substance.

16

# Feedback

If you enjoyed teaching TOPS please tell us so. Your praise motivates us to work hard. If you found an error or can suggest ways to improve this module, we need to hear about that too. Your criticism will help us improve our next new edition. Would you like information about our other publications? Ask us to send you our latest catalog free of charge.

For whatever reason, we'd love to hear from you. We include this self-mailer for your convenience.

Sincerely,

*Ron & Peg*

**Ron and Peg Marson**
author and illustrator

## Your Message Here:

Module Title _____ Date _____

Name _____ School _____

Address _____

City _____ State _____ Zip _____

─────────────────────── FIRST FOLD ───────────────────────

─────────────────────── SECOND FOLD ───────────────────────

RETURN ADDRESS

─────────────────────────────
─────────────────────────────
─────────────────────────────

TOPS Learning Systems
342 S Plumas St
Willows, CA 95988

TAPE HERE

www.ingramcontent.com/pod-product-compliance
Lightning Source LLC
Chambersburg PA
CBHW081513200326
41518CB00015B/2493